THE COMMONWEALTH AND INTERNATIONAL LIBRARY
Joint Chairmen of the Honorary Editorial Advisory Board
SIR ROBERT ROBINSON, O.M., F.R.S., LONDON
DEAN ATHELSTAN SPILHAUS, MINNESOTA
Publisher: ROBERT MAXWELL, M.C., M.P.

GEOPHYSICS DIVISION
General Editors: J. A. JACOBS AND J. T. WILSON

An Introduction to Marine Geology

C.S.S. *Hudson*: photograph courtesy of the Department of Energy, Mines and Resources of Canada.

An Introduction to Marine Geology

by

M. J. KEEN

ASSOCIATE PROFESSOR, INSTITUTE OF OCEANOGRAPHY,
DALHOUSIE UNIVERSITY, HALIFAX, NOVA SCOTIA

THE QUEEN'S AWARD
TO INDUSTRY 1966

PERGAMON PRESS

OXFORD · LONDON · EDINBURGH · NEW YORK
TORONTO · SYDNEY · PARIS · BRAUNSCHWEIG

Pergamon Press Ltd., Headington Hill Hall, Oxford
4 & 5 Fitzroy Square, London W.1
Pergamon Press (Scotland) Ltd., 2 & 3 Teviot Place, Edinburgh 1
Pergamon Press Inc., Maxwell House, Fairview Park, Elmsford, New York 10523
Pergamon of Canada Ltd., 207 Queen's Quay West, Toronto 1
Pergamon Press (Aust.) Pty. Ltd., 19a Boundary Street, Rushcutters Bay,
N.S.W. 2011, Australia
Pergamon Press S.A.R.L., 24 rue des Écoles, Paris 5^e
Vieweg & Sohn GmbH, Burgplatz 1, Braunschweig

First edition 1968
Reprinted 1969
Library of Congress Catalog Card No. 67—26689

Printed in Great Britain by Hazell Watson & Viney Ltd., Aylesbury, Bucks

08 012505 0 (flexicover)
08 012506 9 (hard cover)

Contents

Preface

I have attempted to present an account of some aspects of marine
geology and marine geophysics, comprehensible to those at an early
stage in their study of geology, and to scientists who are not specialists
in these fields. There are many who, being biologists, chemists, mathe-
maticians or physicists, work in the laboratory or on board ship with
geologists and geophysicists, and this book may help them to understand
the aims of their colleagues' experiments. I have not hesitated to simplify
terminology where there is no loss of necessary precision. No attempt
has been made to present a full account of all possible aspects of marine
geology that might be considered, and I have not referred to near-shore
processes at all. Many sections are explanatory and will be familiar to
any geologist who by chance reads the book; he should, of course, turn
the pages until he finds something less familiar to him.

Dalhousie University   M. J. KEEN
Halifax, Nova Scotia

Note To First Reprint

Ideas about ocean-floor spreading developed rapidly during the production of this book. The references listed below may help the reader. Some are already referred to, and are collected here for convenience. The reader's attention is also drawn to *The Sea,* Vol. iv, to appear in 1969.

Bullard, E. C., Reversals of the earth's magnetic field, *Phil. Trans. Roy. Soc. Lond. A* **263,** 481 – 524 (1968)

Isacks, B., Oliver, J. and Sykes, L. R., Seismology and the new global tectonics, *Jour. Geophys. Res.* **73,** 5855 – 5900 (1968)

Le Pinchon, X., Viscosity of the mantle from relaxation time spectra of isostatic adjustment, *Jour. Geophys. Res.* **73,** 3661 – 3698 (1968)

Mckenzie, D. P., Some remarks on heat flow and gravity anomalies, *Jour. Geophys. Res.* **72,** 6261 – 6274 (1967)

McKenzie, D. and Parker, R. L., The North Pacific: an example of tectonics on a sphere, *Nature* **216,** 1276 – 1280 (1967)

Morgan, W. J., Rises, trenches, great faults, and crustal blocks, *Jour. Geophys. Res.* **73,** 1959 – 1982 (1968)

Sykes, L. R., Mechanism of earthquakes and nature of faulting on the mid-oceanic ridges, *Jour. Geophys. Res.* **72,** 2131 – 2153 (1967)

Vine, F. J. and Mathews, D. H., Magnetic anomalies over ocean ridges, *Nature* **199,** 947 – 949 (1963)

Wilson, J. Tuzo, A new class of faults and their bearing on continental drift. *Nature* **207,** 343 – 347 (1965)

Acknowledgements

I am grateful to many for help and advice at various times, among them F. Aumento, Miss S. Fullerton, Miss C. Griffin, R. E. Heffler, Miss B. Hendry, F. R. Hayes, K. S. Manchester, J. I. Marlowe, F. Medioli, D. J. Stanley, D. J. P. Swift, P. J. Wangersky, M. J. L. Willett and Miss C. A. Young. J. E. Blanchard, C. M. Boyd, Mrs. C. E. Keen, B. D. Loncarevic, P. E. Schenk and J. Stewart read parts of the manuscript, and D. Simpson, F. J. Vine and J. Tuzo Wilson read all of an earlier version of it; their care and attention is appreciated. I should like also to acknowledge the assistance of the Bedford Institute of Oceanography, the Department of Energy, Mines and Resources of Canada, the National Research Council and the Defence Research Board of Canada in many experiments undertaken by Dalhousie University.

My gratitude and thanks go especially to the late M. N. Hill for encouragement and personal kindness over a number of years.

The following societies, journals, organizations, and publishers very kindly gave permission for reproduction of illustrations: American Association for the Advancement of Science, American Geophysical Union, *Canadian Journal of Earth Sciences*, Clarendon Press, Cushman Foundation for Foraminiferal Research, Department of Mines and Technical Surveys of Canada, Department of Scientific and Industrial Research of New Zealand, Elsevier Publishing Company, John Wiley and Sons, McGraw-Hill Book Company, *Nature*, the Royal Society, Seismological Society of America, Society of Economic Palaeontologists and Mineralogists, and the United States Geological Survey.

CHAPTER 1

Introduction

This book is an account of some of the rocks at the bottom of the sea. Readers who are not geologists may find geological terminology confusing, and in an effort to help them an elementary account of some geological phenomena is presented in this chapter. The appendixes contain brief systematic descriptions of rock-forming minerals and a summary of the geological time-scale.

THE GROSS STRUCTURE OF THE EARTH

The earth is nearly spherical, but it is flattened at the north and south poles, so that the polar radius is about 6357 km and the radius at the equator is about 6378 km. This leads to the length of one degree of latitude changing from the equator to the poles; Newton first suggested that this might be so, and Bouguer and de Maupertuis demonstrated it by measuring the length of one degree of latitude in Peru near the equator, and in Lapland. The mass of the earth is $5 \cdot 977 \times 10^{27}$ g and its mean density is $5 \cdot 517$ g/cm^3; because the densities of most rocks which are found on the surface of the earth are in the range $1 \cdot 5$ to $3 \cdot 5$ g/cm^3, the density of the earth must increase below the surface. The moment of inertia of the earth can be calculated from astronomical observations, and it is found to be less than that of a body the same size, mass and shape as the earth but of uniform density throughout. This indicates that the earth's density must increase towards the centre, and, moreover, it dictates a boundary condition which must be met by any proposed density distribution.[42]

The study of earthquakes and of artificial explosions shows that there are a number of discontinuities within the earth; these are defined initially in terms of the velocity with which a disturbance caused by an earthquake or other source of energy is propagated. This velocity can be related to the physical properties of the medium through which the energy travels; in conjunction with other properties such as the size, mass and moment of inertia, reasonable deductions may be made about

1

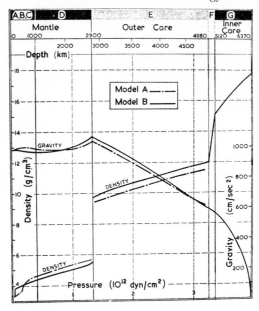

Fig. 1.1. Pressure, density and gravity within the earth for two models which have been proposed (after J. A. Jacobs,[149] from data by K. E. Bullen).

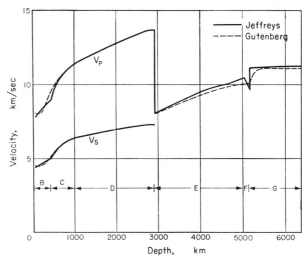

Fig. 1.2. Compressional (V_p) and shear (V_s) wave velocities as a function of depth; models by two scientists are shown (after J. A. Jacobs,[149] from F. Birch).

the density distribution, and one distribution which has been proposed is shown in Fig. 1.1.[41] This figure, and Fig. 1.2, shows the major discontinuities within the earth and the velocities with which compressional and shear waves travel. A number of types of waves may travel through a homogeneous elastic medium. Those associated with the surface of the medium, or with boundaries between different media, are surface waves. Compressional waves are those in which the particle motion within the medium is in the direction of propagation of the wave, and shear waves are those in which the particle motion is perpendicular to the direction of propagation; the velocity of compressional waves is greater than that of shear waves. Shear waves can only travel in a solid.

There are two major discontinuities within the earth; one lies at a depth of approximately 2900 km and the other at a depth which ranges from about 10 km to 70 km. The first is the boundary between the mantle and the core, and the passage from mantle to core is marked by a sharp decrease in the velocity of compressional waves, which suggests that the density increases across the discontinuity. Observations of shear waves suggest that a part of the core is fluid in the sense that shear waves are not propagated through its outer part.

The second discontinuity is the Mohorovičić discontinuity, often written M, and this separates the thin outer part of the earth, called the crust, from the mantle. This is recognized by the change in the velocity with which compressional and shear waves are propagated; it will be discussed in more detail in Chapter 8.

THE ROCKS WHICH FORM THE OUTER PART OF THE EARTH

Any answer given to the question "What is the earth made of?" is unlikely to be wholly correct; our direct evidence of composition is most inadequate, for any sampling of the surface or any drillings a few kilometres into the earth yield evidence of the composition of only a small fraction of the whole mass of the earth. Many statements concerning the composition of the earth must be based upon indirect evidence. Seismological experiments allow deductions to be made which concern the variation in elastic properties of substances at depth; we can infer from experiments in laboratories some of the properties of the rocks which we find on the earth's surface under the conditions which we may suppose exist at depth. Meteorites give evidence of extra-terrestrial materials in the solar system.

There are three major groups of rocks to be observed on the earth's surface. *Sedimentary* rocks include those which are the result of deposition of inorganic and organic particles in water, or result from the aggregation of particles by wind, or are the product of evaporation of saline solutions. Limestones, shales and sandstones are examples of this type of rock. Lava may be seen to issue from a volcano and to solidify on cooling; rocks formed by the cooling and solidification of a liquid which was once at a temperature substantially greater than the normal ambient temperature of the earth's surface are called *igneous* rocks. Many igneous rocks are not, of course, formed at the surface, but are injected at depth. The lava which runs out heats the surface over which it flows, and careful examination of the rocks under the surface will show that they differ from the rocks where no lava has run. Rocks changed in some way from their original state are called *metamorphic* rocks; the change which takes place when the hot lava runs over the cold surface is one example of this metamorphism, called thermal metamorphism. This division of rocks into three categories is somewhat arbitrary, for some will fall partly into one category and partly into another, but it is useful in some ways. Again, it is difficult to put names to rocks which can be always rigidly applied, but the names are useful as handles, as aids to communication. What is important is the significance of the minerals which comprise the rocks and the conditions and processes which led to their formation. Since we can neither give names to rocks nor discuss their particular significance without knowing what they are made of, their components must be considered. These components, or minerals, fall into a number of groups, of which the silicates, oxides and carbonates are examples.

The physical properties of the rock-forming minerals and their arrangement or fabric determine the properties of the rock of which they are part. The rocks are broken down by weathering, and different minerals react in different ways to weathering processes. For example, the clay mineral montmorillonite is one product of the weathering of volcanic rocks in the sea, and the distribution of the clay minerals assumes significance in relation to the distribution of volcanic rocks. Kaolinite is a common product of the weathering of some igneous rocks on the continents, so that its distribution on the sea floor may give a clue to the effects of ocean currents upon the transport of particles.

The rocks of continents are a great mixture of sedimentary, igneous and metamorphic rocks. The major structural elements of each continent are the shields, units which are relatively stable, with which only few earthquakes are associated, and which have formed platforms on and around which younger rocks have been superposed. The rocks of the shields are all Precambrian, ranging in age from between 2000 and 3000 million years old to between 800 and 1000 million years old. They are composed of a variety of rocks of all types, most rather metamorphosed (generally speaking). Their ages have been estimated from studies of radioactive nuclides, and the Canadian Shield appears to decrease in age outwards from a central nucleus, perhaps representing continental growth by accretion of rock at the edges. It has also been suggested that this only reflects successive periods of disturbance – metamorphism and igneous intrusion, and that the same materials are regurgitated from time to time. Rocks are found which can be compared with modern unmetamorphosed equivalents – the lava suites in the Superior Province of the Canadian Shield, for example, although now metamorphosed were once similar to suites of lavas poured out today. Questions arise, then, about the origin of the ancient lavas, whether their source was in the mantle of the earth, and, bearing in mind the predominance of volcanic rocks in the ocean basins, whether the sources of the ancient lavas were similar to those of the lavas of the continents or ocean floors of today.

The rocks of the mountain ranges most familiar to us – the Rocky Mountains, the Appalachians, the Andes and the Alps among them – are less metamorphosed than the remnants of mountain ranges of the shields, and their structural features more easily distinguished. Not all mountain ranges are alike – the Himalayas occupy a position between two parts of a continent now united, whereas the Rocky Mountains and the Andes are at the western margin of North America. The suggestion arises that the Himalayas were generated by the coalescence of two continents, and this implies that at least one of them sailed, as it were, across an ocean basin. Evidence might be sought then, not only for this drift of continents but also for large displacements and fractures of the sea floor. If continents drift then their latitudes may have changed over geological time and their climates altered. Climate will also change with extensive flooding of the continents. Consequently fauna and flora found fossilized in the sediments of the continents should change, not only in an evolutionary manner with time, but with change in environment. Studies of the fauna and flora of the seas at the present time, and of the physical conditions under which they live, will enable a better reconstruction to be made of the conditions under which fossils lived. Similarly, sedimentary rocks have many analogues in modern sediments; the distribution of particular minerals over the floors of the oceans depends upon their proximity to continents and upon ocean currents. The interpretation of the conditions under which fossils lived, and the conditions under which sedimentary rocks were formed, is not as easy as was once thought. Formerly an investigator might have said that an increase in grain size in a sediment in some particular direction implied a source in that direction; but the effect of ocean currents is known now to be a most important factor affecting the distribution patterns within sediments.

The topography of the sea floor is different from that of the continents, and the structure of the oceanic crust is different from the structure of the continental crust. The internal structure of the crust is known not directly through drilling, except in a few isolated places so far, but indirectly through measurement of physical properties. The knowledge gained this way forms the whole basis for our understanding of the ocean basins, and the next chapter is an account of the physical methods which have been used and the intrepretation of the measurements made. The remaining part of this chapter is a review of some rock-forming minerals and some igneous rocks. It will be familiar material to any geologist.

ROCK-FORMING MINERALS

Silicates

Silicates are minerals formed largely of silicon, oxygen and metal atoms. The silicon and oxygen atoms are arranged in the form of a tetrahedron, with one silicon atom at the centre of four oxygen atoms. A number of other metal atoms can also, being small enough, occupy the silicon site, aluminium and germanium among them. The tetrahedra may be isolated one from the other in the sense that no oxygen atoms of adjacent tetrahedra are shared, or they may be linked by sharing of common oxygen atoms. The arrangement of the tetrahedra is a convenient means for describing and classifying the silicates, and the different arrangements lead to different physical properties. The structural arrangements recognized are (1) isolated tetrahedra, (2) double tetrahedra, (3) rings, (4) single chains, (5) double chains, (6) sheets and (7) framework structures.

The isolated tetrahedra have as their basic unit one silicon atom and four oxygen atoms $(SiO_4)^{4-}$. The valency electrons are taken up by metal ions which, if they are the right size, can fit between the individual tetrahedra. Magnesium and ferrous iron are both suitable and the *olivines* are the group of minerals with the general formula $(Mg,Fe)_2SiO_4$. Mg^{2+} and Fe^{2+} are similar in size and consequently there is a complete atomic substitution series from Mg_2SiO_4 to Fe_2SiO_4. By contrast, Ca^{2+} is larger than Mg^{2+} and Fe^{2+}, and Ca_2SiO_4 does not form a part of the series. The tetrahedra in the olivines are arranged compactly, which leads to relatively high densities (Mg_2SiO_4, $3 \cdot 22$ g/cm³) and the occurrence of olivines as equi-dimensional crystals, not platy (like micas) or fibrous (like amphiboles). Another possible linkage of independent tetrahedra yields the garnets, which are also found as equi-dimensional crystals. They have the general formula $R_3^{2+}R_2^{3+}Si_3O_{12}$, where R^{2+} and R^{3+} are metal cations, such as Ca^{2+}, Fe^{2+}, Mg^{2+}, and Mn^{2+}, and Al^{3+}, Fe^{3+} and Cr^{3+}. Bonding of tetrahedra by shared oxygen atoms is stronger than the bonding by metal cations which is found in the olivines. Consequently the olivines weather more easily than quartz, for example a framework silicate in which all tetrahedra are linked by oxygen atoms. In their turn feldspars weather more easily than quartz, because although they are also framework silicates, the lattice is distorted and weakened by incorporation of metal cations. The difference in bonding leads also to differences in such properties as cleavage; pyroxenes (see below) cleave easily by breaking bonds between chains or rods of tetrahedra, the chains being linked one to the other by metal cations.

Two oxygen atoms from each tetrahedron may be shared with another tetrahedron, linked fore and aft; in this case we will have as a basic unit two silicon atoms and six oxygen atoms, $(Si_2O_6)^{4-}$. Calcium, magnesium and ferrous iron are the principal metal ions that fit between the tetrahedra, and lead to a group of minerals with general formula (Ca, Mg, Fe)$_2Si_2O_6$, the *pyroxenes*, in which the tetrahedra are arranged in single chains. The chains may be linked together in pairs, resulting in the structural formula $(Si_4O_{11})^{8-}$ and the remaining ions are both metal ions such as Ca^{2+}, Mg^{2+}, Fe^{2+}, for example, and hydroxyl ions $(OH)^-$. Among groups with this double chain structural arrangement, and reflecting this arrangement, often a fibrous habit, are the *amphiboles*. The most common of these is the mineral hornblende. This is rather complex, and a less common member of the amphiboles, tremolite, is better as an example of the group. It has the formula $(OH)_2Ca_2Mg_5(Si_4O_{11})_2$. The logical extension of these structural configurations is to form an array of tetrahedra infinite in two dimensions, that is, only one oxygen atom in each tetrahedron is not shared by any other; sheets will be formed and we will have a basic unit $(Si_4O_{10})_2^{8-}$. Four hydroxyl groups $(OH)^-$ occur with each unit and are arranged in the part of the layer which contains the oxygen atoms that are not shared. This leads to $[(Si_4O_{10})_2(OH)_4]^{12-}$, but because aluminium may substitute for silicon in the tetrahedra we will have not this but $[(Al_2Si_6)O_{20}(OH)_4]^{14-}$. The metal ions Mg^{2+}, Fe^{2+}, Al^{3+}, Na^{2+}, K^+ fit in between the sheets. Among mineral groups of this type are the *micas*, and whilst the formula may be more complex in practice, an ideal *muscovite* mica will be $(Al_2Si_6)O_{20}(OH)_4K_2Al_4$. The micas are sheet-like in habit, after their structural arrangement. Only one oxygen in the mica group is not shared with a second SiO_4 tetrahedron, and if this is shared a three-dimensional array of oxygen and silicon atoms is formed. The groups of minerals with a structure of this sort include the feldspars and quartz. The more common feldspars are of two types, the plagioclase feldspars ($NaAlSi_3O_8$, $CaAl_2Si_2O_8$) and the alkali feldspars ($K, Na)AlSi_3O_8$. The silica group has the general formula SiO_2, and the feldspars have a similar formula except that either one or two aluminium atoms substitutes for one or two silicon atoms in every four.

There are a number of polymorphs of silica, e.g. quartz, tridymite and cristobalite. These differ in the arrangement of the tetrahedra, and are stable under different conditions of pressure and temperature. Their structures decrease in compactness from quartz to cristobalite, and consequently the densities of the minerals decrease in the same order.

A number of variants in the structural arrangements are possible. For example, two tetrahedra may be joined and this leads to the structural unit $[Si_2O_7]^{6-}$; an example of a mineral group in which this is the structural unit are the *melilites*. Rings may be formed by association of six tetrahedra as $[Si_6O_{18}]^{12-}$, and *beryl* is a typical example of such a ring-type mineral. The influence of the structural arrangement upon the physical properties is very great and it is found that in beryl, for example, the ring arrangement leads to large spaces available for atoms such as argon which are not necessary to complete the structure of the mineral.

A summary of the important rock-forming silicates is presented as Appendix 2.

Carbonates

The principal carbonates with which we shall be concerned are calcium carbonate ($CaCO_3$) as the minerals calcite or aragonite, and dolomite $CaMg(CO_3)_2$.

Calcium carbonate may occur in several crystalline arrangements of which the most common are calcite and aragonite. The calcareous tests (or shells, more loosely) of a number of groups of animals such as the Mollusca and the Foraminifera are made of either calcite or aragonite, or a mixture of both; calcite is the more stable form of the two and aragonite transforms into it. The calcium ion may be replaced in part by a number of others, such as Mg^{2+}, Mn^{2+}, Fe^{2+} and Sr^{2+}, and the degree of replacement depends upon factors such as the relative sizes of the ions and the space into which they must fit. Strontium, for example, is more easily accommodated in aragonite, and on conversion of aragonite to calcite some of the strontium may be lost. A study of the strontium content of sediments might therefore be useful in understanding the environment in which the sediment formed. The isotopic composition of the oxygen in calcium carbonate, in particular the ratio O^{18}/O^{16} has been used in attempts to determine the temperature at which the carbonate formed, and such studies will be considered in more detail in a later section.

Dolomite $(CaMg(CO_3)_2)$ forms usually through alteration of calcite and aragonite by addition of magnesium without disturbance of the CO_3^{2-} ion (see Chapter 4). However, Arrhenius[7] has pointed out that dolomite crystals thought on account of their shape to have formed *in situ* have been observed in recent deep-sea sediments, and that calcium carbonate may be "dolomitized" by basaltic intrusions. Dolomitization of sediments may occur in two different ways—as alteration either of unconsolidated calcium carbonate-bearing sediment or of lithified (turned into rock) sediment.

IGNEOUS ROCKS

Rocks such as solidified lava flows, which are the product of crystallization of a hot molten mass, are called igneous rocks. Igneous rocks may be broadly defined as rocks which were once rather mobile and were probably hot. This is somewhat vague, but the point is not too important; the difference between the modes of formation of a limestone which is composed of dead coral, and a basalt formed of solidified lava is clear. If we are concerned with a particular granite or a particular metamorphic rock, the difference may not be so clear, but we have then to discuss them in detail, and the general classification becomes unimportant.

The properties of a melt or magma depend upon the chemical composition and the physical conditions such as temperature and pressure. Higher pressure increases tne concentration of volatile phases that can be contained in the melt, and this has the effect of decreasing the melting point of the solid phase which may form. A decrease in pressure can cause crystallization just as cooling can.

Molten volcanic materials are erupted at temperatures between 800 and 1200°C at the surface of the earth as lavas, and on eruption the surface of the molten flowing lava is often chilled sufficiently quickly so that a glass is formed rather than discrete mineral grains. As the lava flows over the ground, the surface beneath is baked. This is obvious but important, for these features may be diagnostic of the origin of the igneous rock. If sediment bounds the igneous rock above and below then we can tell by the baking of the sediment whether or not the rock intruded into the sediment as a whole or flowed over the lower layer of sediment as a lava flow. The fabric of the volcanic rock (i.e. the arrangement of its components, crystals or assemblages of crystals) may be diagnostic of the prevailing conditions at the time of solidification. For example, there may be isolated, relatively large crystals of one or more minerals set in a fine-grained ground mass, giving the rock a porphyritic structure. These large crystals, or phenocrysts, may reveal details of the process of crystallization of the melt, and of the overall composition of the source of the lava. For example, if crystals of olivine form the phenocrysts, then it is important to note whether or not they show signs of re-sorption, or of reaction at their edges. If they do, they may not have been in equilibrium with the melt, and may be a product of early crystallization or of accumulation from elsewhere. Lavas may have cavities in them called vesicles, attributed to escaping gases. These may be filled in subsequently by various minerals, such as zeolites. Their presence or absence may be important when investigating lavas poured out on the sea floor; high hydrostatic pressure might prevent vesicle formation.

Rocks which cool slowly by comparison with volcanic rocks, and at some depth within the earth, are called plutonic. Several factors will control the rate of cooling; among them are the depth at which the body is located and the surrounding temperature, the nature of the container of the melt and the size of the molten body. Thin sheet-like bodies may be rather fine-grained, indicating rapid cooling, whereas large bodies may be coarse-grained, which will suggest slow cooling.

The forms which plutonic rocks take are many. Among them are the following:

Sills. Tabular bodies concordant with the major structure of the invaded rock, such as bedding and foliation.

Dykes. Tabular, often vertical or steeply dipping sheets, cutting across the trend of structure of the invaded rocks.

Ring dykes. Steeply dipping dykes of arcuate or annular outcrop.

Batholiths. Large intrusive bodies with walls which dip steeply and with bases which are not seen.

Plutons. A general term for all intrusive bodies, especially if the form falls into no well-defined category.

A CLASSIFICATION OF IGNEOUS ROCKS

The purpose of a classification is to enable readers to have the same opinion as others have about words such as "granite" and "basalt" when they are used. Our need is for a handle so that instead of saying "coarse-grained igneous rock containing quartz (20%), orthoclase (70%), muscovite (10%)", we say "granite".

Igneous rocks can be divided into a number of groups based upon their mineralogy. At one end of the grouping are the rocks relatively rich in the mineral quartz (SiO_2), and in those minerals for which the SiO_2 percentage (of a chemical analysis) is high—the feldspars, with which are associated micas or hornblendes. When coarse-grained these rocks are called granites and granodiorites (depending upon the proportions of alkali and plagioclase feldspar) and form plutons; when fine-grained they are called rhyolites and dacites, and form thin intrusions (dykes and sills) or extrusions—lava flows. The plagioclase feldspars are sodium rich, not calcium rich. These rocks are termed *acidic*, and the percentage of SiO_2 (the so-called acid-forming radical) in a chemical analysis will be high—near 70%. The acidic rocks are common as large batholiths hundreds of kilometres long and wide on the continents, of which those of British Columbia and south-west England are well-known examples. By contrast, granites and granodiorites are rarely found within the ocean basins proper, for reasons discussed in Chapter 7.

The *intermediate* rocks are less rich in silica, and this is reflected in their mineralogy, for typically they contain little or no quartz. The feldspars are similar to those of the acid rocks. The coarse-grained plutonic rocks are syenites and diorites, the fine-grained equivalents trachytes and andesites. They are, in general, less common on the continents than either the acidic rocks or the basic rocks, gabbros and basalts which reflects their probable origin. They are, however, found in large quantity around the margin of the Pacific Ocean (hence the *andesite line*).

The basic rocks are less rich still in silica, and they are composed of a mixture of the mineral groups olivines and pyroxenes (one or both), with calcium-rich plagioclase feldspar. Gabbros are coarse-grained, basalts fine-grained. Basic rocks are found as intrusions or lava flows on the continents and make up much of the ocean floor and oceanic islands. The low silica content is reflected typically by the presence of silica-poor minerals (such as olivines) and the absence of quartz. Ultrabasic rocks, relatively very poor in SiO_2, are discussed in Chapter 7. They consist predominantly of one or more of olivine and pyroxene.

It is easy to become familiar with the grouping, and with the mineralogy and chemistry of the igneous rocks, if the intimate dependence of each upon the other is appreciated. Figure 1.3 is an attempt to put the various factors together in a very rough and schematic way.

The bottom line of the lower part of the figure has decreasing quartz content to the right, and increasing content of minerals deficient

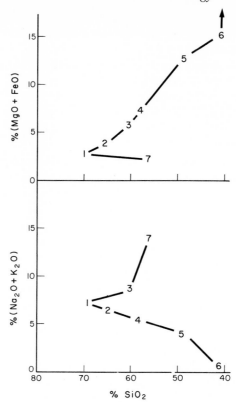

Fig. 1.3. Chemical variations in common igneous rocks. *Upper*: (MgO + FeO) plotted against SiO_2. *Lower*: (Na$_2$O + K$_2$O) plotted against SiO_2. (1) granite, rhyolite; (2) granodiorite, dacite; (3) syenite, trachyte; (4) diorite, andesite; (5) gabbro, basalt; (6) dunite; (7) nepheline–syenite, phonolite. (The first of each pair is coarse-grained, the second the fine-grained equivalent.)

in SiO_2, rich in calcium, magnesium and iron — pyroxenes and olivines. The content of calcium plagioclase increases relative to sodium plagioclase and alkali feldspar (reflecting the decrease in Na$_2$O and K$_2$O). The upper "line" also has decreasing quartz content to the right, but the increase in Na$_2$O and K$_2$O leads to minerals such as nepheline, rather than olivines and pyroxenes. The decrease in (Na$_2$O + K$_2$O) from granites through granodiorites and diorites to dunites is accompanied by increase in (MgO + FeO), corresponding to the increasing content of ferromagnesian minerals, pyroxenes and olivines.

CHAPTER 2

Geophysical Techniques for Exploration of the Sea Floor

INTRODUCTION

One object of marine geology is to describe and account for the rocks that compose the sea floor and the physical and chemical processes to which they are subject. We may study rocks on land near the surface of the earth by more-or-less direct examination; rocks at great depth cannot be so studied because of their inaccessibility. Inacessibility is the major difficulty of marine geology. To overcome it there have been developed a number of methods which depend upon the measurement of physical properties and interpretation of such measurements. We can, for example, study the velocity of propagation of elastic waves generated by an artificial explosion or earthquake or the variation in the value of the acceleration due to gravity. The interpretation of such measurements will seldom yield a unique solution but a combination of a number of physical measurements may lead to one which, whilst not unique, is more likely than most.

If a rock, or the water which overlies the rock, is sharply struck, as happens in an explosion or earthquake, the disturbance so generated is propagated away from the source of the disturbance. The manner in which it is so propagated may be characteristic of one or more types of rock, under the particular conditions of temperature and pressure prevailing. Methods of investigation concerned with these phenomena are those of seismology. One rock type may be more dense than another, and arranged as a body of particular shape in a particular position within the earth. The acceleration due to gravity depends upon the density and distribution of masses of rock, and measurements of the acceleration may lead to the establishment of models of the distribution of bodies of different densities which would lead to the observed values of the acceleration. Similarly, the magnetic field of the earth is due in part to the magnetization of the rocks in the outermost few tens of kilometres, and measurement of the strength or intensity of the field and its direction may lead to various distributions of rock bodies with particular magnetic properties being postulated to satisfy the field which is actually observed. The amount of heat which flows out of the earth through a known area in a known time will depend upon the sources of heat within the earth; sources of heat include among them radioactive processes and a study of heat-flow over the earth will enable restrictions to be placed upon the nature and distribution of such sources. Some rocks are more radio-active than others and heat-flow measurements may therefore let us say something about the distribution of particular types of rock.

EXPLOSION SEISMOLOGY

The methods of explosion seismology are those in which the energy transmitted through water and rock generated by an artificial source, such as an explosion or an electric arc discharge, is observed. We may be concerned with the observation of reflections from a discontinuity, such as that between water and rock at the bottom of the sea, or that between two bodies made of rocks with different elastic properties; observation of the first discontinuity mentioned is of course "echo-sounding". Observation of the reflections from the second type of discontinuity may be called "reflection-profiling". There are paths of propagation of the disturbance away from the source other than reflection paths— energy may be transmitted in a way such that the time which a pulse takes to travel corresponds to a ray path, the ray being refracted in the manner of geometrical optics. Methods of investigation using such paths are those of "seismic refraction". Whatever the method, sources and detectors are required.

Sources and Receivers

The sources of energy which may be used are numerous and depend upon the particular purpose. In echo-sounding a high-frequency pulse is transmitted to the water using perhaps a piezoelectric or magneto-strictive transducer to convert electrical into elastic energy which is propagated through the water. The frequency transmitted will be 10 kc/s, or higher. Reflection-profiling (to observe the geological structure beneath the sea floor) may be carried out with small explosive charges, or with sparks between electrodes in the water. Explosives may be used in seismic refraction studies, the amount of charge depending on its type and the nature of the work. At ranges between source and receiver of 50 km or more, charges greater than 100 kg will be required.

The disturbance which arrives at the recording station, a change in pressure or displacement, is detected by transducers, geophones or hydrophones. Geophones respond to ground motion. A magnet is suspended from a spring inside a coil, and displacements of the ground cause relative motion between magnet and coil so that an e.m.f. is generated. This signal can be amplified and undesirable frequencies filtered out. The characteristics of geophones vary with application, but in long range refraction studies at ranges of tens to hundreds of kilometres the dominant frequencies recorded are in the range of 3–20 c/s and geophones are used which have resonant frequencies of 2 c/s or lower. Hydrophones respond to changes in pressure. A diaphragm may convert pressure changes to mechanical motions detected in a way similar to that in the geophone described, or piezoelectric transducers may convert the changes in pressure directly into electrical signals.

The electrical output which results is amplified—filtered if necessary and recorded—on strip chart recorders (photographically or by pen), magnetic tape, or on "graphic" recorders. (On a graphic recorder the time of arrival of a sound pulse is indicated by the position of a mark on the recording paper; the intensity of the mark depends upon the intensity of the sound pulse.)

REFLECTIONS: ECHO-SOUNDING AND SEISMIC-PROFILING

Suppose a sound pulse is generated by a source (which acts also as detector) which is a distance d away from a reflector. The pulse travels along the path from source to reflector to detector (at the same place as the source). If the velocity of sound transmission is v then the time of transmission from source back to source is $2d/v$. The time can be measured and if the value of v is known d can be calculated. Echo-sounding is an example of this exercise in timing (Fig. 2.1), and a number of difficulties arise.

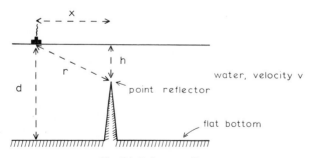

Fig. 2.1. Echo-sounding.

The sound pulse will not be transmitted as a narrow parallel beam but more nearly as a cone. The analysis below shows that this will lead to a point reflector appearing on the record as an hyperbola. The angle of the cone may be approximately 20°, so that in water 4 km deep the cone intersects the ocean floor in a circle of diameter 1·4 km, and the lateral resolution is thereby limited. The vertical resolution is more precise, for it depends more nearly only on the precision of timing, and on the vertical motion of the ship; however, echoes from reflectors not in the vertical plane which contains the ship's course, but which are at a lesser distance from the ship than reflectors beneath it, will produce side echoes and confuse the vertical resolution. The sound pulse has a finite wavelength, which is 15 cm at a frequency of 10 kc/s and objects not greater in dimension than several wavelengths cannot be resolved. The ship will be displaced vertically by the sea through some metres, and this will obviously limit the vertical resolution.

Isolated peaks may act as point reflectors. If, as in Fig. 2.1, an isolated peak which is at a depth h is in the field of the transducer, then reflections from the top of the peak can be recorded first so long as

$$(x^2 + h^2)^{\frac{1}{2}} < d,$$

where d is the depth to the (flat) bottom of the sea. The travel time ship-to-peak-to-ship is t where

$$t = 2(x^2 + h^2)^{\frac{1}{2}} /v.$$

If t is plotted against x with constant h and v, an hyperbola results and this is often seen on recordings of soundings in areas where the topography is rough and peaks are jagged. Consequently, sounders of narrow beam are very desirable.

The true depth can be found only if the velocity is known. The velocity of sound in sea water varies with pressure, temperature and salinity so variations throughout the oceans are considerable, and either the velocity or the physical conditions must be measured. Tabulations exist which allow approximate corrections to be made.[196] Confusion in reporting soundings is immense for we have at our disposal many ways of so doing: assume the velocity of sound propagation in water is 800 fm/s or 1500 m/s, and do not correct for variations from either of these values: assume the corrections of Matthews' tables: assume some other corrections. Soundings are still on the published charts which were obtained by lead-line sounding, especially in remote areas, which adds to the confusion.

The principles of seismic-profiling are similar to those of echo-sounding, but the frequencies generated and recorded are lower. Lower frequencies are recorded because high-frequency sounds are attenuated in rock more rapidly. The intention of experiments in seismic-profiling is to map reflecting horizons in sediments, or the interface between sediments and igneous rocks. The velocity of propagation in the whole section above the reflector must be known if the true thickness of the section is to be deduced. Although it is possible to obtain values of velocity from observations of reflections away from normal incidence, often experiments of a different kind are made to obtain this information on velocity.

Seismic Refraction Experiments

The principles. Seismic reflection experiments are concerned with the reflection of energy at an interface, with the travel time between source and detector, and with the amplitude and frequency of the reflected wave. Seismic refraction experiments differ from reflection experiments in the path of propagation: the energy we are concerned with is that which appears in terms of the travel time to be refracted at an interface between one medium and another, as is shown in Fig. 2.2. The information available is the shape of the signal received at the detector, and the time that has elapsed between the instant the energy left the source and

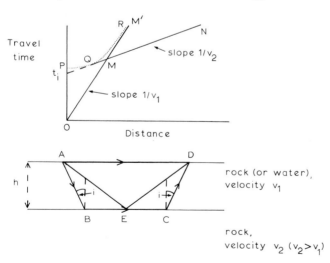

Fig. 2.2. Seismic refraction experiments: a simple model and the corresponding time–distance curve.

the instant that energy arrived at the detector. Most studies made at sea have taken only the travel time into account; recent studies on land have been as much concerned with the information contained in the signal shape as with the travel times. Consideration of the results of interpretation suggest that both time and distance should be measured with an accuracy near 0·1%. Travel times range up to a few minutes in many experiments, and the measurement of time must be to within 0·01 sec or better.

The plot of travel times against distance is often of the form shown in Fig. 2.2, which shows also one model of geological structure which could satisfy the observations. Suppose the shot point at A and the detector D are on a horizontal surface and are a distance x apart. Consider a horizontal layer in which the velocity of propagation is v_1 (compressional or shear, as we please, at this moment), and this layer overlies a medium in which the velocity is v_2, this velocity being greater than v_1. The media are homogeneous and isotropic and the boundary is horizontal and sharp.

The time for energy to travel along the path AD directly is t_1, where

$$t_1 = x/v_1. \tag{1}$$

The time for energy to travel the path $ABCD$ is t_2, where

$$t_2 = AB/v_1 + BC/v_2 + CD/v_1,$$

that is,

$$t_2 = x/v_2 + 2h/v_1 \cos i - 2h \tan i/v_2. \tag{2}$$

The angle of incidence at B and the angle of refraction at C is i: the thickness of the upper layer is h. The angle of incidence i is derived from Snell's Law, which applied at B or C becomes

$$\sin i/v_1 = \sin 90°/v_2$$

or

$$\sin i = v_1/v_2.$$

By substitution, equation (2) becomes

$$t_2 = x/v_2 + 2h\sqrt{(v_2^2 - v_1^2)}/v_1 v_2. \tag{3}$$

If values of t are plotted against values of x straight lines will be obtained, and the values of the reciprocals of the slopes will be the velocities of propagation. In Fig. 2.2 OM' corresponds to energy which has travelled directly along AD and the reciprocal of the slope of OM is v_1; MN corresponds to energy which has travelled along $ABCD$ and the reciprocal of the slope is v_2.

The thickness h can now be found. If MN is projected back to intersect the ordinate we obtain the time intercept t_i corresponding to the value of t for x equal to zero in equation (3); then h can be found, for

$$t_i = 2h\sqrt{(v_2^2 - v_1^2)}/v_1 v_2. \tag{4}$$

This type of model can be extended to any number of layers and to layers which are not horizontal but which dip. If any layer is not horizontal then it is not sufficient to obtain the plot of travel times which results from firing charges in one direction only; a reversed profile must be obtained, one in which charges are fired between detectors placed at both ends of the profile (at A and D in Fig. 2.2).

It may be more convenient and there is only a small loss in rigour if an end-to-end profile is obtained; in this case the detectors are in the centre of a line of charges fired either side of the detectors.

The justification for the use of models of this nature is that the "refraction arrivals" have been shown to exist in model experiments and can be predicted theoretically.[32,82] It may be in some circumstances more realistic to suppose that there is a continuous velocity gradient rather than a series of discrete layers in which the velocity is uniform.[258-9] Velocity increases with pressure and decreases with temperature; both temperature and pressure increase with depth and the exact form assumed by the change in velocity depends critically upon the temperature gradient; it is not easy to predict this. Birch has shown by experiment[19-20] how the velocity of propagation in a number of rock types changes with pressure; there is a rapid increase in velocity in the range of pressure which corresponds in depth to the upper 5 km even if any reasonable temperature gradient is taken into account. Below 5 km the change in velocity is less rapid (Fig. 8.15). The time–distance curve that would result from a change in velocity with depth of this type may be little different from a straight line. A number of possible models which would satisfy observed time–distance curves is shown in Fig. 2.3.

The models proposed must be tested then on other evidence; this could consist of travel time data for paths other than those of AD and $ABCD$ of Fig. 2.2, or of studies of the signal shapes of the phases received at the detectors. Figure 2.2 shows another possible path, and the time–distance plot that corresponds to it. OM' and OMN correspond to the paths AD and $ABCD$ respectively; PQR corresponds to the reflection path

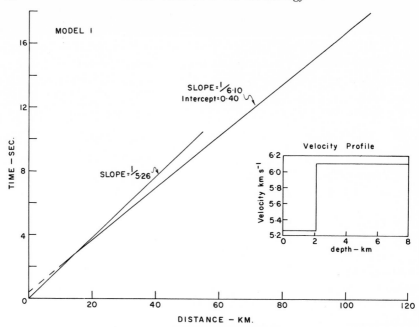

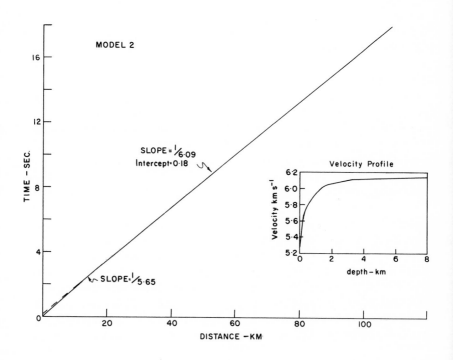

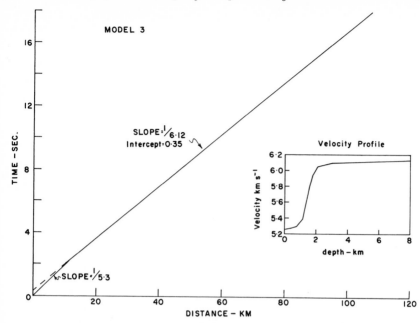

Fig. 2.3. Time–distance curves which correspond to velocity gradients (shown inset).

AED, and this curve is tangent to *QN* at the point *Q* corresponding to the closest distance at which refraction at the critical angle is possible. At large values of *x* the times for the reflection and for the direct wave approach each other, and so *PQR* is asymptotic to *OM'*. One test of the model will consist of testing for the presence of reflected phases at the appropriate time. To do this well is difficult, and the test must be done well to distinguish the arrival of reflected phases from the arrival of phases that may have travelled along other paths that are possible.

Sediment thickness in the deep-sea basins. The basis for interpretation of models which consist of homogeneous layers is that the velocity increases in each succeeding lower layer. If a layer is sufficiently thick then at some range between shot and receiver the energy which has travelled horizontally through this layer will be the first to arrive at the receiver, and these successive "first arrivals" will lie on a straight line. If the layer is not thick enough or the differences between the velocity in the layer and the velocities in the layers above and below not great enough, then the layer will not be represented by first arrivals and it may be difficult to identify the layer in the analysis of the results. This is illustrated in Fig. 2.4, which shows the time–distance curve corresponding to one very simple model of the oceanic crust. The velocity of sound in water is about 1·5 km/s and the compressional wave velocity in the main crustal layer of the ocean basin is about 6·7 km/s. Thin layers of rock such as uncon-

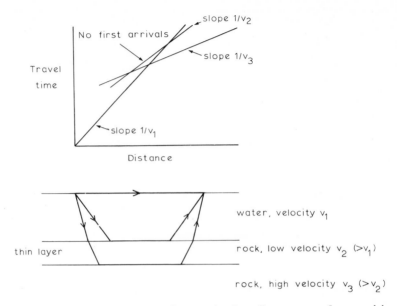

Fig. 2.4. Seismic refraction experiments: the time–distance curve for a model with a relatively thin, relatively low-velocity layer.

solidated sediment or consolidated sediment with compressional wave velocity between 1·5 and 6·7 km/s, may not yield first arrivals on a time–distance curve if shots are fired and arrivals are received at the surface of the ocean. For this reason the thickness of sediments on the ocean floor was not at one time well determined, but the use of detectors on the ocean floor, reflection techniques and gravity measurements have overcome the difficulty.

SEISMOLOGICAL INVESTIGATIONS: EARTHQUAKES

Compressional and Shear Waves

Compressional (*P*) and shear waves (*S*) are generated by an explosion or by an earthquake. *P* waves can travel through solids and liquids but

S waves can travel only through solids. Their velocity depends on the elastic constants and the density of the medium through which they travel and they are only slightly dispersed, that is, their velocity is nearly (but not quite) independent of frequency.[91,307] If the times of arrival of P and S are determined at a number of observatories for a number of earthquakes then we may calculate the distribution of velocity with depth within the earth. This leads to some knowledge of the elastic constants at depth and makes possible at least tentative identification of the rock types within the earth. Difficulties arise in determination of the structure of the crust and the upper part of the mantle because the travel times are not known with sufficient certainty and because the distribution of seismological observatories over the surface of the earth has been uneven and in many parts of the world has been very poor. The distribution of observatories throughout the oceanic regions has been, for obvious reasons, most inadequate. As is the case with the interpretation of seismic refraction explosion studies, the solution of problems in which there exists a low-velocity layer is hard.

In recent years there have been major advances in our knowledge of the structure of the crust and upper mantle of the earth for several reasons. First, the world seismological observatory network has been extended. Second, large arrays of detectors have been constructed; these arrays, such as that at Yellowknife, Northwest Territories, Canada, extend for tens of kilometres, and act as highly directive antennae for elastic waves.[313] Third, experiments have been conducted with explosions large enough to be of value in investigations not only of the earth's crust, but also of the upper part of the mantle of the earth.[17]

Surface Waves

As their name implies surface waves travel at the boundary of an elastic medium, and they are of two main types, Love waves and Rayleigh waves. The particle motion in Love waves is parallel to the plane of the surface and perpendicular to the direction of propagation, whereas in Rayleigh waves the motion is in the vertical plane and retrograde and elliptical along the direction of propagation. They are both dispersive, that is their velocity of propagation is frequency-dependent, and the velocity is a function of the elastic properties over the depth of penetration; this depth is approximately one wavelength.

The dispersion of surface waves has been used to deduce crustal structure along paths from earthquakes to observatories which are wholly continental or wholly oceanic and such studies are most useful in complementing explosion refraction studies. They have been of especial value in investigation of the structure of the upper part of the mantle where near-earthquake studies are inadequate. Miss Lehmann[170] suggested that a low velocity zone existed in the upper few hundred kilometres of the mantle, and this was supported by studies of Gutenberg.[104] Surface wave analysis has confirmed this suggestion and further has shown that the depth to the low velocity zone is different under continents and oceans.[60]

GRAVITY AT SEA

Introduction

The attraction between two masses M_1, M_2 at a distance R apart, which is so great that the masses may be considered as point objects, is

$$F = G(M_1 M_2/R^2).$$

G is the gravitational constant. Consequently, the acceleration of M_1 towards M_2 is

$$f = G(M_2/R^2).$$

Were the earth exactly spherical and either uniformly dense throughout or composed of spherical shells each of uniform density, then the acceleration at the surface of an earth of radius R and mass M would be

$$g = G(M/R^2).$$

If the values of the quantities M and R suggested in Chapter 1 are substituted in this formula for g, and G is taken to be $6 \cdot 67 \times 10^{-8}$ c.g.s.u., then g is found to be rather less than 1000 cm/s², the precise figure depending upon the value of R which is taken.

The earth is not exactly spherical, or uniformly dense or composed of uniformly dense shells. The value of the acceleration g towards the centre of the earth is not constant over the surface of the earth, and the direction of the acceleration towards the centre of the earth does not coincide with the direction which is normal to the earth's surface. The values of g over the earth's surface may depend upon the variation in density throughout the earth, and any distributions proposed must lead to the observed values of g.

It will be seen that it is easier to measure changes in the values of g than to measure the absolute value of g itself. This absolute value is not known at any point on the earth's surface with the accuracy to which changes can be measured and one value has been adopted as a base value, which will be rather close to the exact value.[49,50] The variations in g are expressed relative to those values which would be obtained upon a particular geometrical figure, an ellipsoid, which coincides approximately with sea-level over the oceans and with the geometrical projection of sea-level over the continents. The difference between the value of g measured and the value on this ideal reference ellipsoid is the *gravity anomaly*. The anomaly may be expressed in various ways. If the value of g which would have been observed had the observation been at sea-level is calculated, the difference is the *free-air anomaly*. In this case, only the effect of the height of the observation above or below sea-level is taken into account. If the gravitational effect of the infinite sheet of rock, of thickness equal to the height of the observation above sea-level, is subtracted from the free-air anomaly a *simple Bouguer anomaly* results. Topographic corrections might also be subtracted if the variations in topography from a plane surface are to be taken into account. The effect of known variations in subsurface density might also be considered, in which case a modified Bouguer anomaly is calculated. A third type of anomaly can also be calculated—the *isostatic anomaly*. Suppose that although there are variations in elevation (of topography) and of density of rocks over the surface, nevertheless at some level below the earth's surface, the compensation level, the masses per unit area above are equal,

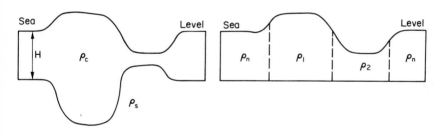

Fig. 2.5. Isostasy: the assumptions of Airy (left) and Pratt. Note that ρ_c (ρ is density) is constant above the lower boundary of the Airy figure, but ρ_1, ρ_2, ρ_n are different from each other above the lower boundary of the Pratt figure (after G. D. Garland[92]).

regardless of location. This can be visualized in two ways (Fig. 2.5).[92] Light crustal material of uniform density may project downwards into a dense substratum wherever the surface is elevated. This is Airy's hypothesis. Alternatively different columns may reach the same level below sea-level but have different densities—Pratt's hypothesis. The attraction of the configuration of masses of either hypothesis may be calculated and subtracted from the Bouguer anomaly, and the result is the isostatic anomaly. The isostatic anomalies have minimum values, in the least-squares sense, if the level of compensation on the Pratt hypothesis is 113·7 km, and between 20—40 km on the Airy hypothesis. This does not imply that either is correct. However, the structure beneath the Puerto Rico trench, to be discussed in Chapter 8, shows non-isostatic equilibrium in the Airy sense. The trench is deeper than the surrounding ocean floor. Consequently, the dense mantle should bulge upwards beneath the trench to account for the bulge downwards of (light) water. In fact the crust beneath the trench thickens. Consequently, at least to relatively shallow depths within the mantle, there is not uniformity of mass per unit area.

The anomalies discussed are fictitious: they simply indicate departures of observed values from those calculated on the basis of an assumed model of distribution of masses. As the model approximation approaches the real geological situation the anomalies should decrease in amplitude. Unfortunately zero differences between calculated and observed anomalies do not mean necessarily that the model assumed is correct.

Acceleration is expressed as centimetres per second per second; it is convenient to use the gal, $1·0$ cm/s²; hence one milligal, $0·001$ cm/s², is approximately one-millionth of the acceleration due to gravity on the surface of the earth, and we shall see that we require measurements at sea to be precise to within one or two milligals.

Measurement

The absolute value of the acceleration due to gravity g, is hard to measure. It is simpler, although still difficult, to measure changes in g from place to place, and a number of methods have been used to do this at sea, of which two are especially important. These are to measure the change in period of a pendulum, and to measure the change in length of a loaded spring.

If we have a simple pendulum of length L the period T_1 with which it will swing if the acceleration due to gravity is g_1 is

$$T_1 = 2\pi\sqrt{(L/g_1)}.$$

The pendulum may be swung at a second place where the acceleration due to gravity is g_2; the period is now T_2 where

$$T_2 = 2\pi\sqrt{(L/g_2)}.$$

The change in g, $(g_2 - g_1)$, can be calculated if T_1 and T_2 are known, and L remains invariant between the measurements, for

$$g_2 - g_1 = g_1 \, (T_1^2/T_2^2 - 1).$$

Care in measurement is needed; effects such as the buoyancy of the air, the flexure of the pendula, the change in length of the pendula with temperature, and others must be taken into account. Descriptions of measurements made in this and other ways have been given by B. D. Loncarevic.[173-4] The procedure is not altered whatever the particular design of pendula or the particular experimental procedure; the same set of pendula is used from place to place and the period is measured.

A second method is to measure the change in length of a spring which supports a mass. The changes in g may be found by determining the changes in the extension. This is the principle used in many gravity meters; care has to be taken to eliminate errors such as those which arise through change of length of the spring with temperature, for example.

Particular difficulties arise at sea because the sea is in motion and because navigation may be imprecise. A gravity meter cannot distinguish between forces caused by the sea and the force due to gravity; the vertical accelerations caused by the sea may range up to 100,000 mgals and the precision with which changes in g should be determined is about 1 mgal. The distribution of energy in ocean waves is such that most is in those with periods between 3 and 18 sec. The motion of the water decreases with depth beneath the sea surface and the motions with short periods are attenuated more rapidly than those with long periods. Such considerations lead to the measurement of gravity with pendula of short period swung in a submarine, and to the use on a surface ship of an instrument the motion of which is damped so that it does not respond to motions of short period. In both cases the measurement must be made over a long period of time; a ship on the surface remains there so that the average value of the vertical accelerations not due to gravity should tend to zero over a sufficiently long interval of time. The instrument may be mounted on a platform stabilized to follow the true vertical or mounted in gimbals. If it is mounted in gimbals then the instrument will follow the apparent vertical, the direction of the resultant of the vertical accelerations and the horizontal accelerations. In this case the average value of the acceleration measured over a reasonable period of time tends towards the acceleration due to gravity plus a correction term. The correction term depends upon the squares of the horizontal accelerations, and does not tend to zero as the averaging time increases.[34]

The value of g depends upon the distribution of mass beneath the vessel carrying the gravity meter and this depends upon the water depth and the geological structure. If the vessel is moving, as a surface ship will be, to measure g over a period of several minutes means that the changes in g due to structures with horizontal dimensions greater than 1 or 2 km will be resolved.

Navigation

Imprecise navigation leads to two difficulties. The gravity field of the earth depends upon the latitude because the earth is flattened and because it rotates. If a value of g is obtained at any particular place it must be compared with the value which would be obtained if the earth were a homogeneous body the figure of which were exactly known. The variation of this value with latitude is approximately 1 mgal per minute of latitude; an error of 1 nautical mile in position in a north–south direction leads to an error of 1 mgal in the theoretical value of g. The theoretical value with which the measured value will be compared depends also upon the velocity of the ship which is making the measurement. This is because the measurement is made upon a vehicle moving in a rotating co-ordinate system. If the ship is moving at constant speed the correction which has to be applied—the Eötvös correction—is approximately 4·04 cos (latitude) mgals per km/hr west to east velocity. Consequently the component of the ship's velocity in this direction must be known within ·2 km/hr if the correction is to be less than 1 mgal.

THE TOTAL INTENSITY OF THE MAGNETIC FIELD
OF THE EARTH

The Field

The earth's magnetic field, measured at the surface or above the surface, is roughly like that which would arise if a bar magnet were located at the centre of the earth, and aligned close to the polar axis. That is, it is a disturbed dipole field.[3, 40] The perturbations of the dipole field fall into two groups according to the scale of their linear dimensions.

One group has dimensions of approximately 1000 km and has its origin in the mantle or core. These perturbations change with a time scale of several hundred years, and their pattern drifts westwards over the surface of the earth. These time variations are the secular variation, and the changes in intensity are in the range of several gammas or tens of gammas per year (1 γ is 10^{-5} oersted). Consequently in surveys of the magnetic field at sea made from year to year they must be taken into account. The other group of perturbations or anomalies has linear dimensions of the order of tens of kilometres or less and can be ascribed to causes within the outer tens of kilometres of the earth. The division is fundamental, for those which have their origin deep within the earth cannot be due to changes in the induced or permanent magnetization of rocks because of the temperature; this is likely to be well above the Curie point of any known magnetic minerals. By contrast, the anomalies in magnetic field with dimensions of the order of tens of kilometres must be due to changes in the magnetization of the rocks which compose the crust and upper part of the mantle. Studies of these anomalies yield information on the geological structure.

The greater part of the magnetic field of the earth is due to internal causes, but small time-dependent fields are generated externally and are indirectly of solar and lunar origin. These fields include among them the diurnal or daily variation, and if we are interested in the spatial variation of the magnetic field these time-dependent fields must either be negligible or corrections for them should be made. The magnitude of the total field ranges from values of about 30,000 γ at the magnetic equator to about 60,000 γ at the magnetic poles, and the magnitude of the spatial anomalies of geological origin range from some tens of gamma to some thousands of gamma. The daily variation is about 50 or 100 γ in range, so that if there is interest in geological anomalies of this order of magnitude the daily variation must be taken into account. In practice, the number of magnetic observatories in oceanic regions being limited and the sea being rather wide, diurnal variation has more often than not been neglected and account taken only of magnetic storms, during which time variations in the field may be large. An interesting development which arose out of a desire to know the diurnal variation when making magnetic surveys at sea is the diurnal variation buoy developed by M. N. Hill and C. S. Mason.[138] Because sea water has high electrical conductivity the changes in the magnetic field with a period of less than one day are attenuated in the central parts of the oceans. Near the continental margins, however, there might be some enhancement of daily variation, as first observed by Hill and Mason.

Measurement of the Field

The total field of the earth has been measured in surveys at sea in two ways, by means of the flux-gate magnetometer and the nuclear precession magnetometer.[136,250,293] The flux-gate magnetometer is the saturable core type in which a rod of ferromagnetic material with non-linear magnetization properties acts as the core of one or more windings. The windings are connected to alternating-current circuits. Changes in the magnetic field cause changes in the magnetization of the core and this is measured by the change in flux produced by the alternating-current. Such a magnetometer can be towed behind a ship or mounted in an aircraft, but it has the disadvantage that it is not very stable and must be calibrated frequently.

The nuclear precession magnetometer is now more commonly used in surveys at sea, and was first developed for this by M. N. Hill.[136] A coil is wound around a bottle containing a liquid rich in protons, such as water. A magnetic field of about 100 oersteds is induced in the liquid by passing a direct current through the coil. This field is reduced to zero by switching off the direct or polarizing current, and the nuclear moments precess about the earth's magnetic field, inducing an e.m.f. in the same coil. The frequency of this signal (but not the amplitude) depends only on the magnitude of the field in which the precession occurs, and is approximately 2000 c/s. It is independent of the orientation of the field so that if the frequency can be measured with sufficient accuracy the field can be determined. In practice, although the field can be measured to 1:50,000 easily, the daily variation and other sources of error determine the precision of a survey. The method has the great advantage that it is absolute, unlike the flux-gate method. The magnetometer must be towed behind the ship a distance equal to about twice the length of the ship so that the field of the ship is not a disturbing influence. This field depends upon the orientation of the ship with respect to the earth's field, and a heading correction must be found before all surveys.

We shall be concerned mainly with results of measurements of the total field and not with the magnitude of the horizontal and vertical components, or with the azimuth of the field. However, all the elements of the field have been measured in world-wide airborne surveys and some of the results of such surveys will be mentioned; the flux-gate instruments for such surveys have been described by P. H. Serson.[250]

Analysis of Total Field Surveys

We will assume that effects such as the daily variation or the heading correction have been removed; the analysis of the survey can then proceed in two steps. The first is to remove trends in the field which are not wanted, and this depends on the scale of the features under investigation. In all cases it is likely that trends with linear dimensions of the order of 1000 km will not be necessary, and can be removed by eye or analytically. If we are interested in geological features with dimensions of only a few kilometres it may be sensible to remove trends due to features on a scale of some tens of kilometres and greater as well. A sophisticated method for removing trends has been described by Fraser Grant.[102] The second step is to determine the geological body which may account for the remaining anomaly.

This step is inherently difficult. As is the case with gravity anomalies there can be no unique solution, although some solutions will be more likely than others. We have to postulate a body of some particular shape with a magnetization of a specified magnitude and direction. The magnetization may be that induced in the rock by the earth's field or be the remanent magnetization of the rock. The remanent magnetization is that acquired by the rock at some former time and may be different in magnitude and direction from the induced magnetization. The magnetic field which such a body would produce is calculated and compared with the field which has been observed, and the parameters defining the body are altered until the calculated and observed fields are the same.[334]

Sometimes an elaborate and time-consuming analysis is unnecessary and valuable information can be obtained by qualitative examination of the results. For example, if no magnetic anomaly is associated with a seamount then it is likely that the seamount is made of a rather non-magnetic rock such as a limestone; if, on the other hand, there is a large magnetic anomaly it is likely that the seamount is made of a magnetic rock such as a basalt. This is a rather inexpensive way of distinguishing between the two possibilities.[23]

MEASUREMENT OF HEAT-FLOW

Measurements in bore-holes on the continents have shown that there is an outward heat-flow of about 1 μcal/cm^2 sec through the outer 1 or 2 km of the earth's surface. It was reasonable to suppose that a large part of this was due to heat generated during radioactive decay within the continental crust. Seismic measurements at sea suggested that the oceanic crust is very different from the continental crust and would be composed principally of rocks in which the content of radioactive elements is small by comparison with those which compose the continental crust. It was thought likely that heat-flow through the oceanic crust would be much smaller than that through the continental crust.[38] This has proved to be untrue.

Two measurements must be made so that the heat-flow may be calculated, the temperature gradient and the thermal conductivity of the medium in which this has been measured. The temperature gradient has been measured with thermistors mounted in out-riggers from the core-barrel of a corer; two or three thermistors are used down the length of the core-barrel and the temperature differences are measured by a bridge enclosed in a pressure container housed in the main body of the corer itself.[38] This method has the advantage that thermal equilibrium is established only a few minutes after the corer has penetrated the sediment, and the thermal conductivity may be found from the core of sediment obtained at the same time. E. C. Bullard originally used thermocouples in an oil-filled probe 4·5 m long and it was necessary to leave the probe in the sediment for 30 min after penetration to allow heat generated to be dissipated. This had the result that the probe was frequently bent on extraction from the sediment. A separate core was obtained as close to the site of the probe as possible so that the conductivity of the sediments could be determined, but except by chance the two sites would be at least more than 1 or 2 km apart.

Determinations of heat-flow on land have to be made in holes at least 400 m deep so that the effect of seasonal thermal fluctuations are avoided, and measurements in the deep-seas have the advantage that such seasonal fluctuations do not occur. A change in temperature of the water at the ocean-bottom of more that 1 or 2°C in times of approximately 1000 years is not easy to imagine and such changes would be necessary if the present heat-flow measured at the sea floor were to be due to temperature changes of the sea in the past. Bullard[38] has shown that for heat-flow as measured now to change by 50% any one of the following temperature changes would be necessary: 0·80°C 50 years ago, 3·6°C 1000 years ago, 11°C 10,000 years ago. The values of heat-flow found in practice range in magnitude from about 0·2 to 8·0 μcal/cm^2 sec and this range cannot be ascribed to different changes in the temperature of the bottom water in different parts of the world. However, R. P. Von Herzen[329] has detected adiabatic and super-adiabatic temperature gradients in the water near the bottom at some stations where heat-flow measurements have been made, and if these gradients are quasi-stable periodic temperature changes of several hundredths °C could arise. Such an effect could explain the greater variability of values of heat-flow found with relatively short probes compared with values found with relatively long probes. A sudden blanketing of the sea floor with sediments of low thermal conductivity will disturb the temperature gradient and lead to apparently low values of heat-flow until equilibrium is re-established. Under conditions of steady sedimentation the effect would not be noticeable until rates of approximately 100 cm per thousand years are reached,[330] much greater than normal values of sedimentation rates in the ocean basins. Nevertheless, in areas where slumps occur the effect could be important. Irregularities in bottom topography will cause minor differences in the measured values of heat-flow from the true values. Measurements in sediment-filled basins of the Atlantic Ocean show less variability than those in basins of the Pacific Ocean, and this could be due to a greater thickness of sediment over-lying igneous rocks in the one region by comparison with the other.

A number of other factors might lead to the measured heat-flow being erroneously ascribed to causes underlying the sediment. The tempera-ture gradient might lead to convection of the water in the sediment, and heat would be generated by metabolizing organisms. The gradient would have to be rather large for the first effect, and the organic content of typi-cal oceanic sediment is too low for the second effect to be important.[38]

Errors in the actual measurement of heat-flow arise in the estimation of the temperature gradient and in measurement of conductivity. Von Herzen and Langseth[330] estimate that in areas of average geothermal gradient ($0 \cdot 06°C/m$), where the probe penetrates sediment in which the conductivity is relatively uniform, these errors are about 10%. A rather precise estimate of errors has been obtained from measurements in the Indian Ocean. Measurements have been made with corers on which three out-rigged probes are mounted each separated by a vertical distance of 4 m. This allows two estimates of heat-flow to be made, from two pairs of probes, the middle one being common. The mean of the deviation from zero difference in values upper pair to lower pair is 7%. Measurements in a drill-hole over an interval of 154 m showed that heat-flow was constant with depth. Consequently the figure of 7% probably represents a reasonable estimate of the error with sensors 4 m apart, or equivalently 5% for sensors 8 m apart.

Values of heat-flow throughout the world have been analysed in a variety of ways; average values in areas with dimensions of 5° of latitude and longitude have been found, for example, and harmonic analysis applied to the data. Von Herzen and Langseth suggest that the variability of results from the same general area is too great for relatively sophisti-cated analyses to have meaning at the present time.

The number of measurements of heat-flow on the continents is less than that for the ocean basins because of the greater difficulty and ex-pense of measurement. J. S. Steinhart has recently attempted to make measurements in lakes by recording temperature changes in the sediment over periods of several months. Measurement over a long time interval will allow the effects of short period fluctuations in temperature to be removed. Such an approach would have great application in waters relatively cold the whole year, as in deep lakes and fjords of parts of Canada.

OTHER METHODS AND DEVELOPMENTS

Time-varying magnetic fields were treated cavalierly as a source of noise to be removed in measurements of the total intensity of the earth's magnetic field. In fact, analysis of the fields, or of induced currents within the earth, leads to information about the electrical conductivity within the earth, because of the dependence of the fields observed at the earth's surface upon the conductivity of the medium beneath.[45,150,165] The particular relevance of such measurements to us is that the oceanic mantle may differ from the continental mantle, and quantitative estimates of the difference in electrical properties can be deduced from observations of the time-varying magnetic fields.

Theories to account for the origin of the earth's surface features are legion. Many assume motion of the crust and mantle. Such theories can only become quantitative if the forces which act upon the crust and the reaction of the crust and mantle to the forces are studied. Measurement of stress within the earth's crust assumes therefore particular importance,[36] and so does measurement and analysis of the tidal deformation of the solid earth.

The analysis of travel times in seismic experiments presented earlier in this chapter is very elementary. A more sophisticated method of analysis is to reduce the data obtained from a number of recording stations by the method of least-squares.[17,333] This has not been applied to experiments in the ocean basins because any one experiment has not been on a sufficiently large scale, but the approach could be most profitable, and especially useful in looking for anisotropy in the elastic properties of crust and mantle. Detection of signals either from earthquakes or from artificial explosions is made more sensitive by the use of arrays of detectors—designed for maximum reception as a radio receiver's antenna may be.[313] There are many limitations to the analysis of travel times. More profitable may be synthesis of the signals themselves, following the lines suggested in a classical experiment by Officer.[219–20] Seismic experiments on land have shown that deep reflections can be obtained fairly easily from near to or at the base of the continental crust. Such experiments would be most profitable if successful at sea.

CHAPTER 3

The Topography of the
Ocean Floor

The topography of the ocean floor is known imperfectly, but as a framework for discussion it can probably be divided without incorrect simplification into the continental margin, the ocean basin floor and the mid-ocean ridge system. Physiographic divisions of the floor of the North Atlantic are shown in Fig. 3.1.

THE CONTINENTAL MARGIN

The continental shelf, slope and rise divide the continents proper from the ocean basins. The precise details of their features vary, but in general the shelf has a low average gradient close to 1:1000, the slope is by contrast steep, with gradients which vary from near vertical to about 1:40 and the gradient of the continental rise at the foot of the slope

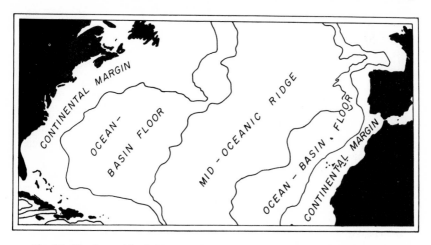

Fig. 3.1. Physiographic divisions of the floor of the North Atlantic Ocean (after B. C. Heezen and H. W. Menard[126]).

35

may be 1:300. The continental shelf is often divided sharply at its edge (the shelf-break) from the slope; the continental rise, where present, merges gently into the abyssal plain at its foot. The type of margin that may occur depends upon the region; for example, the continental shelf off the eastern seaboard of the United States is relatively

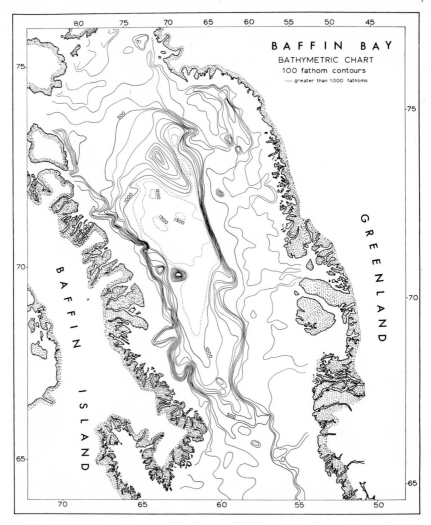

Fig. 3.2. Bathymetric chart of Baffin Bay, between Greenland and Canada. The sill to the south is Davis Strait, where the water depth is approximately 300 fathoms. Note the wide shelf on the east of the Bay, and the narrow shelf on the west of the Bay (compiled by K. S. Manchester, from published sources and from precision soundings made by C.C.G.S. *Labrador*).

wide, whereas off California it is narrow. The first of these areas is rather stable, and by comparison with the second, aseismic. Similarly, the continental slope off the east coast of North America merges into the continental rise and subsequently into the abyssal plain, whereas off the continental margins of a substantial fraction of the Pacific coast of South America and part of North America are trenches, the continental rise being absent.[280] The break in slope at the edge of the shelf varies in

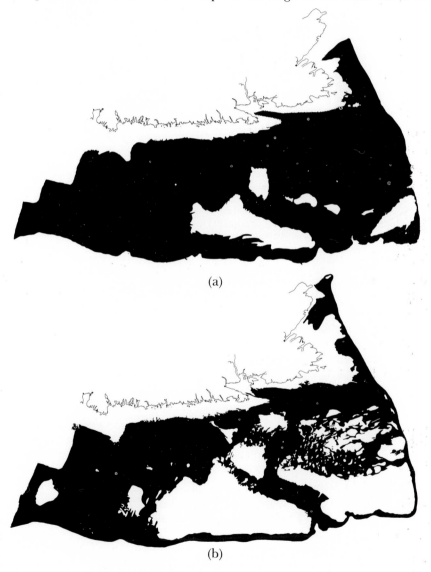

(a)

(b)

(c)

Fig. 3.3. The continental shelf off Nova Scotia (see Fig. 3.6 for location). Presentation to show the complexity of the topography and the distribution of land and sea as it would appear if sea level were reduced successively. The land is white, the sea black (but note that the area beyond the outer black margin is wholly sea). (a) Sea level at the present 30 fm line, (b) 50 fm, (c) 70 fm. The Gully of Fig. 3.6 appears conspicuously in (b) and (c). Note the complexity of the eastern part of this shelf (after soundings by the Canadian Hydrographic Service, prepared by A. E. Cok).

depth from a few metres to several hundreds of metres; off Nova Scotia, where the shelf is 200 km wide the break is at a depth of about 160 m, whilst off west Greenland in Baffin Bay it is at a depth of 800 m[48,332] (Fig. 3.2). The morphology of the shelf depends too upon the underlying rocks, the nature of the erosional or depositional processes and the structural features such as faults. It may be most complicated, as is that off Nova Scotia, shown in Fig. 3.3. The continental shelves of regions of high northern latitudes are often complicated by trough-like features some hundreds of metres deep which may run either perpendicular or parallel to the edge of the shelf.[142] Shelf-like features may be physically separated from the major part of the shelf proper, as is the Flemish Cap off Newfoundland, and "Galicia" Bank, which rises to some 600 m below sea-level off Cape Finisterre on the Iberian Peninsula, separated from the peninsula by water depths of approximately 2500 m.[23] The Blake Plateau is a similarly anomalous feature,[126] called a "marginal plateau" (Fig. 8.21). This plateau has in a sense two continental slopes, the one from the shelf off the continent and the other which leads down to the Hatteras abyssal plain.

The shapes that may be assumed by the continental slope and the continental rise are varied too. The slope off Baffin Island is smooth and gentle and merges imperceptibly with the abyssal plain, whilst that off south-west Greenland is rugged in character and has a sort of marginal trench near its foot. Profiles across the continental margin off the eastern seaboard of the United States, from Cape Hatteras, Virginia, to the Blake Plateau, are shown in Fig. 3.4.[129] We see that there seems to be a

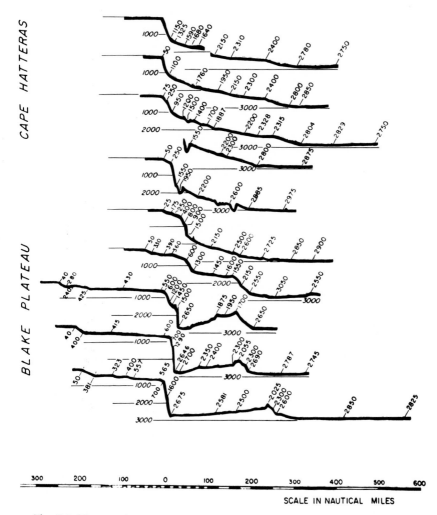

Fig. 3.4. The continental margin off eastern North America: bathymetric profiles east of Cape Hatteras, Virginia, and the Blake Plateau (after B. C. Heezen and H. W. Menard[126]).

development of an outer ridge and a marginal escarpment at the foot of the slope in the southerly profiles. The comment has been made that the lower part of the continental rise off New England can be traced to the Outer Ridge of the Antilles, and that whereas off the Antilles a trench such as the Puerto Rico trench lies between the land and the outer ridge, off New England the trench has been filled in by sediment.[126] Similar features are seen off California and Mexico[280] (Fig. 3.5).

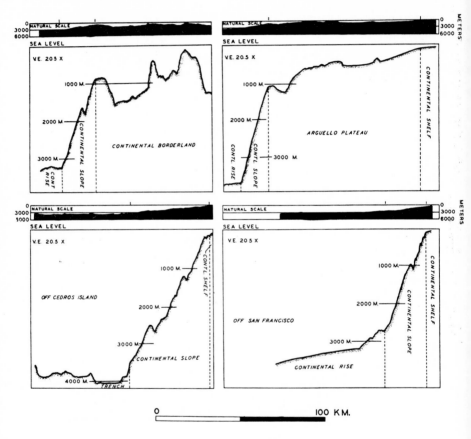

Fig. 3.5. The continental margin off Mexico: typical profiles (after E. Uchupi and K. O. Emery[280]).

The edges of the continental shelves and the continental slopes are cut in places by canyons, called *submarine canyons*. These may be V-shaped and several kilometres in width at the top of the V, up to hundreds of metres in depth and may have steep-sided walls.[88] An example of one from the continental margin of the western part of the North Atlantic, the Gully, is shown in Figs. 3.6 and 3.7. The difficulties which have been found in accounting for the origins of these canyons will be referred to subsequently, but we may note from Figs. 3.6 and 3.8 that the head of

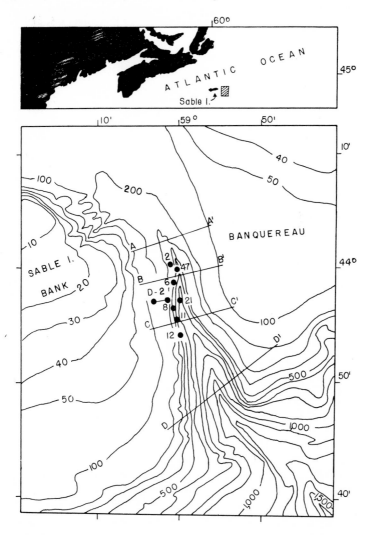

Fig. 3.6. The Gully, a submarine canyon east of Sable Island. Depths in fathoms, corrected. The locations of the profiles of Fig. 3.7. are shown (after J. I. Marlowe[188]).

this canyon off Nova Scotia is well defined just east of Sable Island. No large river mouth is at its head, although, at a time of lower sea-level it could have acted as a drainage channel from the inner region of the shelf (Fig. 3.3). A source of supply for sediment is at present available from the banks along the outer margin of the shelf.

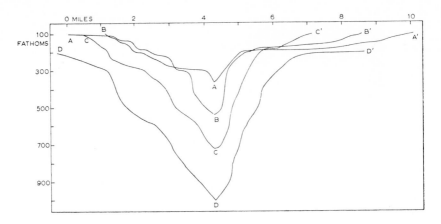

Fig. 3.7. Transverse profiles across The Gully. Lowest points in profiles adjusted to a common axis. Locations of the profiles are given in Fig. 3.6. (after J. I. Marlowe[188]).

At the foot of submarine canyons may be found deep-sea fans and abyssal cones, piles of sediment transported down the canyons. The fans may be cut by channels, sometimes levéed, which lead across the continental rise. If the continental rise is absent, and the margin is bordered by a trench, then the trench may possess an abyssal plain at its foot. The existence of marginal trenches in the Pacific Ocean in greater abundance than in the Atlantic Ocean has the consequence that sediment from the continents is trapped near the borders of the Pacific Ocean, and does not reach the ocean floor beyond the trenches. In the Atlantic this is not so, and abyssal plains in the Atlantic on the ocean floor are more common than in the Pacific.

The margins of the Pacific Ocean are noteworthy for the profusion of deep trenches in the sea floor, associated with island arcs on the side nearest the continent (e.g. the Aleutian trench and Aleutian Islands), or with a mountain range if the trench is adjacent to a continent (e.g. the Peru–Chile trench and the Andes) (Figs. 3.12, 8.22, 8.23 and 8.24). Although they may be considered to define the edge of the Pacific Ocean basin, in fact some are very far from the nearest continental massif – the Marianas arc is 2000 km east of Asia. The earth's crust on the continental side of island arcs is not in any way a continental crust (see Chapter 8), and the sea between island arc and continent may be as

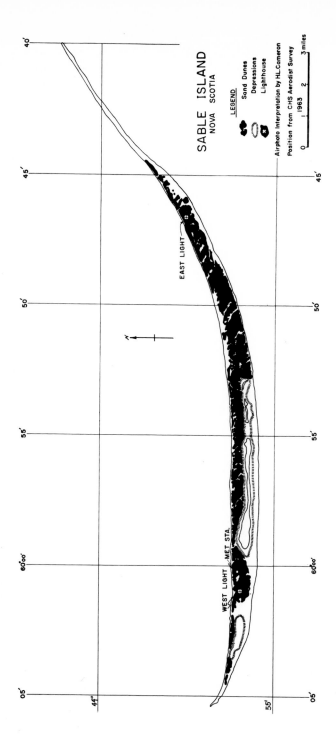

Fig. 3.8. Sable Island, near the edge of the continental shelf off Nova Scotia. The island is made largely of sand dunes, stabilized by a palaeosol[197] (after J. E. Berger, from a survey by the Canadian Hydrographic Service).

deep as the main ocean basin.[(199)] Island arc systems with associated
trenches are found too in the Antillean arc of the Caribbean and the
Scotian arc of the South Atlantic. These are in a sense adjacent to the
Pacific Ocean where the American continent is narrow, or absent. The
absence of arcs and trenches should be noted too. There is no trench off
western North America between the Gulf of California and Alaska,
which coincides with the juxtaposition of crest or flank of a mid-ocean
ridge, the East Pacific Rise, with North America.

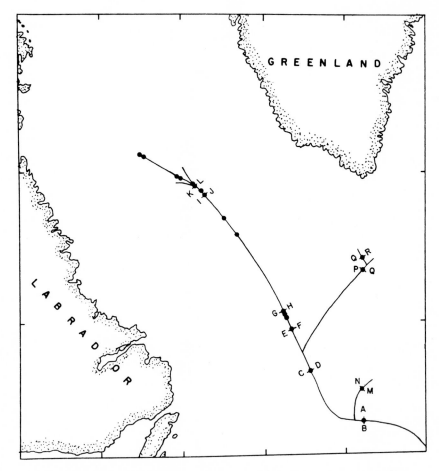

Fig. 3.9. The Northwest Atlantic Mid-Ocean Canyon: location. The letters show
the location of the profiles of Fig. 3.10 (after precision soundings from C.S.S.
Baffin, by K. S. Manchester).

THE OCEAN BASIN FLOOR

The ocean basin floors occupy the regions between the continental margins and the mid-ocean ridges, and have a variety of topographic forms within them, some more-or-less unrelated structurally one to another.

At the foot of the continental rises, and on a smaller scale at the bottom of trenches, lie abyssal plains; their chief characteristic is that they are

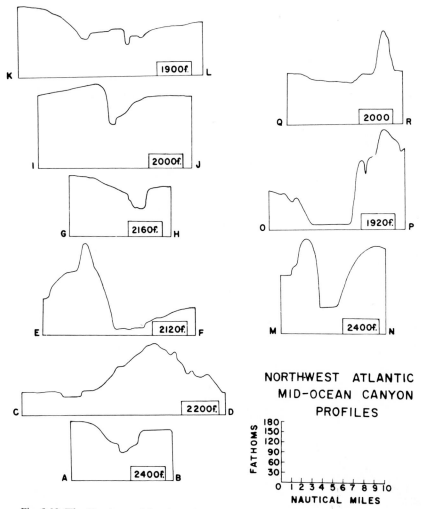

Fig. 3.10. The Northwest Atlantic Mid-Ocean Canyon: bathymetric profiles. For location see Fig. 3.9 (after precision soundings from C.S.S. *Baffin*, by K.S. Manchester).

flat, with gradients of less than 1:1000. Two off western Europe are shown in Fig. 5.1, and a contour map of the central part of the western margin of the Iberian plain in Fig. 5.4. Their flatness has been ascribed to sediments originating on adjacent topographic rises and spread by "turbidity currents" across the floor. Adjacent abyssal plains may be separated by sills, and an excellent example of this phenomenon, where the sill has been cut by an interplain channel so that the two plains are linked, is Theta Gap, west of the northern part of the Iberian Peninsula. This gap joins the Biscay plain and the Iberian plain; it is shown in Fig. 5.3.

The abyssal plains may be cut not only by channels which lead from submarine canyons, but also by *mid-ocean canyons*, of which a fine example is the Northwest Atlantic Mid-Ocean Canyon, found in the Labrador Sea and which leads through a gap into the Sohm abyssal plain. This is shown in Figs. 3.9, 3.10 and 3.11. The origin of this, as of submarine canyons, is not obvious; the first suggestion one might make is that it is cut by sediments transported as turbidity currents, originating in the main part from the Davis Strait and Hudson Strait. This view

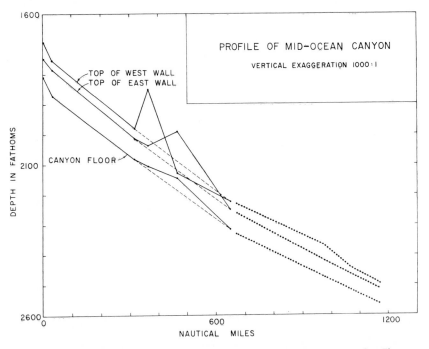

Fig. 3.11. The gradient of the Northwest Atlantic Mid-Ocean canyon. See Figs. 3.9 and 3.10 (after precision soundings from C.S.S. *Baffin*, by K. S. Manchester).

receives support from levée-like forms associated with it, and from the system of tributaries which joins it. The west wall is higher than the east wall. K. S. Manchester has suggested that this can be accounted for by preferential sediment deposition from turbidity currents, caused by the effect of the earth's rotation. The system has not had the attention it deserves.

Abyssal hills rise a few hundred metres above the abyssal plains, or are found at the foot of mid-ocean ridges; examples of those which protude from the Iberian plain are seen in Fig. 5.2. They merge with increasing height into seamounts. Seamounts and associated forms called guyots, the tops of which are nearly flat, occur in large numbers in the Pacific Ocean, where they may form long linear chains with the summits some hundreds of metres below sea-level. It has been suggested that whilst seamounts occur in groups (such as the Kelvin or New England Seamount Chain of the western part of the North Atlantic Ocean) abyssal hills are common over the whole of the ocean basin floor.[126] They may be partially buried by sediment, but are there nevertheless. This indicates that even in areas now aseismic and non-

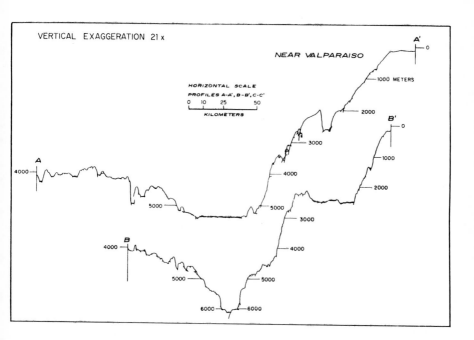

Fig. 3.12. Topography of the Peru–Chile trench. Bathymetric profiles off northern Chile (after R. L. Fisher and R. W. Raitt[87]).

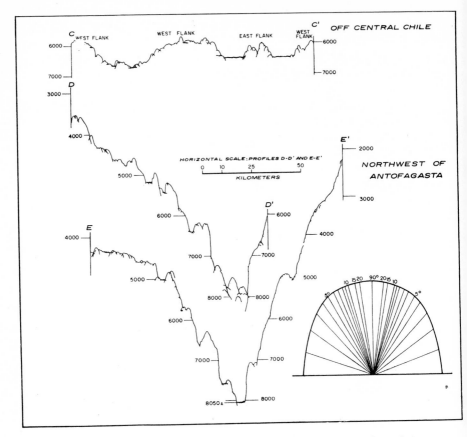

Fig. 3.12. Topography of the Peru–Chile trench. Bathymetric profiles of the trench off northern Chile, traced from precision depth recorder tapes. Sounding scales corrected after Matthews[196]. The rose shows true angles (after R. L. Fisher and R. W. Raitt[87]).

volcanic, a large part of the sea floor has at some time been subject to vulcanism and tectonic activity. This is evident especially on mid-ocean ridges. Fracture zones transect large sections of the floor of the eastern part of the Pacific Ocean, and parts of the Mid-Atlantic Ridge and Carlsberg Ridge.[116,127,193,282] Their existence in the Pacific Ocean is shown by the topography, and the displacements along the fracture zones by studies of the total intensity of the earth's magnetic field.[203,282] The zones in the Pacific are shown in Fig. 6.11. Displacements of the crest of the Mid-Atlantic Ridge are illustrated in Fig. 8.4.

Aseismic ridges may run from the mid-ocean ridges to points on the continents[300] which may, by some hypotheses, have been once joined.

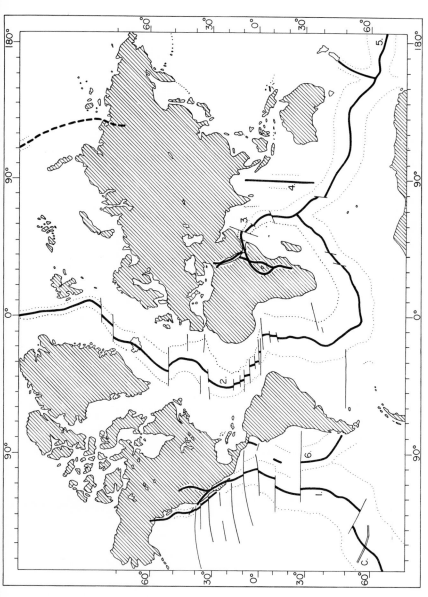

Fig. 3.13. Mid-ocean ridges and rises, (1) East Pacific Rise, (2) Mid-Atlantic Ridge, (3) Carlsberg Ridge, (4) 90° E. Ridge, (5) Pacific–Antarctic Ridge, (6) Galapagos–Chile Rise (after D. H. Matthews[194]).

An example is the Walvis Ridge which runs southwest from Walvis Bay in south-west Africa. The Lomonosov Ridge in the Arctic Ocean is also aseismic; in this case it appears to run between north-west Greenland and Ellesmere Island towards the New Siberian Islands, which lie between the Lartev Shelf and the East Siberia Shelf (see Fig. 8.12). Oceanic rises are large areas elevated above the main ocean floor and may also be aseismic; the Bermuda Rise is an example in the Atlantic Ocean.

MID-OCEAN RIDGE SYSTEM

The mid-ocean ridge is a mountain chain system which extends throughout the world's oceans (Fig. 3.13). In the Atlantic Ocean it is called the Mid-Atlantic Ridge, in the Pacific Ocean it is formed by the East Pacific Rise and the Pacific–Antarctic Ridge, and in the Indian Ocean by the Carlsberg Ridge. The various parts can be defined by their topographic forms and by their geophysical characteristics. These are similar but not identical and Menard[199] has questioned the continuity of the system, preferring to regard the various parts as having a similar underlying cause but being in different stages of development.

The Mid-Atlantic Ridge is, as a topographic feature, about 1000 km wide in places and rises 1 to 3 km or so from the ocean basin floors which flank it to east and west. It has been divided [126] into several topographic provinces; these are the rift mountains and the central rift valley along

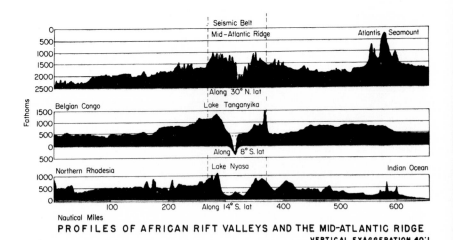

PROFILES OF AFRICAN RIFT VALLEYS AND THE MID-ATLANTIC RIDGE
VERTICAL EXAGGERATION 40:1

Fig. 3.14. Comparison of profiles across East African Rift Valleys and the Mid-Atlantic Ridge (after B. C. Heezen and H. W. Menard [126]).

the crest of the ridge, a high fractured plateau and various steps which form the flanks of the ridge. These features are shown in Fig. 3.14, where the ridge is compared with rift valleys of East Africa, and in Fig. 3.15. The most striking of these features is the central rift valley,

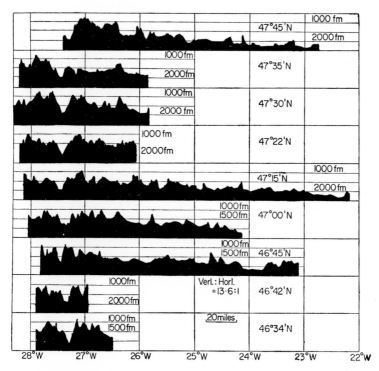

Fig. 3.15. East–west bathymetric profiles from the eastern flanks to the centre of the Mid-Atlantic Ridge. The median rift valley is the conspicuous notch between 27°W. and 28°W. (after M. N. Hill[137]).

the floor of which lies 1000 to 3000 m below the mountain peaks on either side. It may in places be blocked,[137] or be discontinuous because of displacement along a fracture zone. Fracture zones across the Mid-Atlantic Ridge are found in the equatorial Atlantic; one is topographically defined by the Romanche trench.[116] A similar trench coinciding with a displacement of the Carlsberg Ridge is the Vema trench in the Indian Ocean.[127] The crest of the Mid-Atlantic Ridge is often characterized by an anomaly in the total intensity of the earth's magnetic field (Figs.8.6 and 8.7), and by high values of heat-flow.[38] The crest is also seismically active,[105,244] and this is seen on a map which shows earthquake epicentres, as in Fig. 8.11 (see Chapter 8).

Many of these characteristics are common to the mid-ocean ridge in other oceans, and the distribution of earthquake epicentres has been used to locate the path taken by a ridge system before it was well known topographically.[121] There are puzzles. For example, the East Pacific Rise, defined topographically and by the values of heat-flow, runs into the North American continent south of California. The Lomonosov ridge is topographically the dominant mountain chain in the Arctic Ocean basin, yet the seismically active zone lies some way from it. It is possible that this seismic zone is in fact accompanied by a ridge system,[121] the geographical difficulties being compounded by the sparseness of soundings. These features are seen in Figs. 8.12 and 8.13.

THE SIGNIFICANCE OF TOPOGRAPHY

The great attention paid to topography by marine geologists arises, of course, from the relative ease with which it may be discovered, and because it may be the only clue to the underlying geology. The deciphering of the fracture zones of the Pacific Ocean is possible because features are observed which are straight, even across rough topography, and they displace bathymetric contours; the general level one side of the fracture zone may be higher than general level on the other side. The geometry is such that the fracture zone must be vertical, and the displacement mainly horizontal. Vertical displacement leading to apparent lateral displacement can be rejected because it leads to absurd demands upon the dip of layers of rock.[199] The heights of guyots above the ocean floor, the heights of the bases of the coral cappings of atolls, and the attitude of the tops of guyots near trenches led Menard to erect the Darwin Rise and deduce its age relationship with at least one trench. The slope of the ocean floor south of the Aleutian trench led Shor to conclude that the development of the Aleutian arc prevented sediment transport south, and led to in-filling of the Bering Sea.[251]

CHAPTER 4

Pelagic Sediments

INTRODUCTION AND DEFINITIONS

The bathymetric map of the ocean floor shows that the abyssal plains differ from the rest of the floor in their extreme flatness (Figs. 5.1, 5.2). The sediments which are found on the plains and the associated deep-sea fans, cones and channels are derived in part from topographic high places nearby, such as continental margins. These sediments differ from those which accumulate on other parts of the ocean floor because they arise through rather catastrophic processes of deposition. They are the subject of Chapter 5. The sediments which accumulate in many other parts of the ocean basins can be termed *pelagic*, which means "of the open sea" (Fig. 4.1). The purposes of a study of pelagic sediments are numerous, and some are the following:

(1) To see what are the processes of sedimentation in the deep-sea today, and what they have been in the past.

(2) To account for the distribution of elements between the oceans and the continents.

(3) To study the changes in fauna and flora both areally and with time. Such changes are important for their own sake, but also lead to information which may be valuable in determining the past climatic conditions on the earth, the circulation in the ocean and in the atmosphere.

The pelagic sediments may be divided into groups according to their composition and this leads to a scheme of this sort:

Inorganic deposits: The so-called "red clay".

Organic deposits: Calcium carbonate oozes — "Globigerina", pteropod and coccolith oozes, for example. Siliceous oozes — radiolarian and diatom oozes, for example.

These types of deposits will not be considered in the light of a rigid scheme, but in a more general way; we shall discuss the fauna and flora, mineralogy and geochemistry, methods of determining their age and applications of studies to problems of stratigraphy, climatology and palaeo-oceanography.

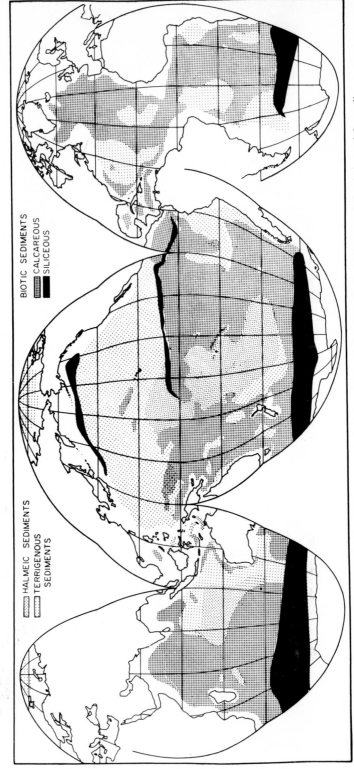

Fig. 4.1. Present distribution of pelagic sediments on the floor of the ocean basins. The term halmeic is used for sediments composed predominantly of minerals precipitated out of solution from sea water (after G. Arrhenius[7]).

HALMEIC SEDIMENTS

TERRIGENOUS
SEDIMENTS

BIOTIC SEDIMENTS

CALCAREOUS

SILICEOUS

FAUNA AND FLORA

The fauna and flora found in pelagic sediments are predominantly the skeletal remains of planktonic and benthonic organisms which are mainly of a calcareous or siliceous composition. Planktonic organisms are organisms which are not capable of strongly directed movement and include among them the Foraminifera and coccolithophorids. Benthonic organisms are those which live at the bottom of the sea. (Readers more familiar with studies of the mud-water interface in shallow environments of continental shelves, lakes, bays and estuaries[112] may be saddened at this cursory treatment of the living fauna and flora. However, deep-ocean cores have been seldom collected in a way which will keep intact the interface, and further, sudden pressure changes break up bacterial cells so that what survives to be cultured may not be representative of the flora.) Among the objectives of a study of the fauna and flora is the determination of the history of the oceans using as a guide the biogenic component of the sediments.[236] The skeletal remains of pelagic organisms which are found (except for whales' ear-bones, for example) are small in size, but may comprise a large fraction of the sediment. The organisms may be divided in the following way, according to the type of skeleton they possess:

Plants
        calcareous: coccolithophorids
        siliceous: diatoms
                         silicoflagellates

Animals
        calcareous: Foraminifera
                         pteropods
        siliceous: Radiolaria
                         sponges
        phosphatic: fish remains

The skeletal remains of these organisms may make up a large part of the sediment; this is true of the Foraminifera, Radiolaria, diatoms and coccolithophorids especially, but the other groups are less common, not only because of their actual abundance in the water but because the skeletons may dissolve too rapidly. The skeletons of pteropods, for example, are made of aragonite and this dissolves more easily than does calcite, with the result that pteropod oozes are less common than foraminiferal oozes and occur in water which is shallow rather than deep.

If we are to use the skeletons of these micro-organisms to determine past oceanographic conditions, then there must be assurance that the present distribution on the ocean floor reflects with sufficient accuracy the distribution in the water masses above. The organisms must also be sensitive to changes in the properties of the water masses, to the temperature of the water at the surface, for example, and the relationship between the changes that are found in the organisms and the properties of the water must be known.

The problem can be attacked in a number of ways. Laboratory studies of the influence of properties such as salinity, temperature, hydrostatic

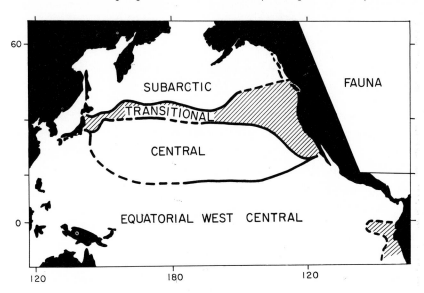

Fig. 4.2. Distribution of living planktonic faunal assemblages in the North Pacific Ocean: general distribution of northern cold-water (subarctic) fauna, southern warm-water fauna (central and equatorial west central) and transition fauna (after J. S. Bradshaw[30]).

pressure and pH are not easy to make because of the difficulty of isolating one factor, say temperature, for study.[29,224] In the case of the diatom *Asterionella japonica* the influence of temperature upon its distribution deduced from experimental studies does not agree with the observed geographical distribution. The influences of environmental factors are not confined to whether or not a factor is lethal or non-lethal, but more important, to their effect upon rate of growth and reproduction. J. S. Bradshaw has found with one foraminiferal species that water temperatures in the range 15–20°C lead to an increase in size, but reproduction takes place only between 18 and 30°C; consequently the presence

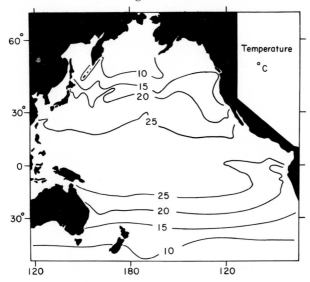

Fig. 4.3. Distribution of sea surface temperatures (°C) in the North Pacific Ocean. (Values obtained at time of collection of samples on which Fig. 4.2 is based, or after H. U. Sverdrup, M. W. Johnson and R. H. Fleming.[261]) (After J. S. Bradshaw.[30]).

of large tests of the species does not necessarily represent optimal environmental conditions but marginal conditions. This observation has its counterpart in the study of Foraminifera in sediment cores. One criterion for recognition of the change from the warmer conditions of the Pliocene to the colder conditions of the Pleistocene is the change in abundance and size of *Globorotalia menardii*. They are larger and less

(a)

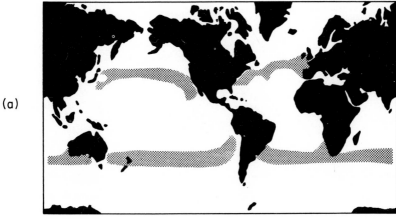

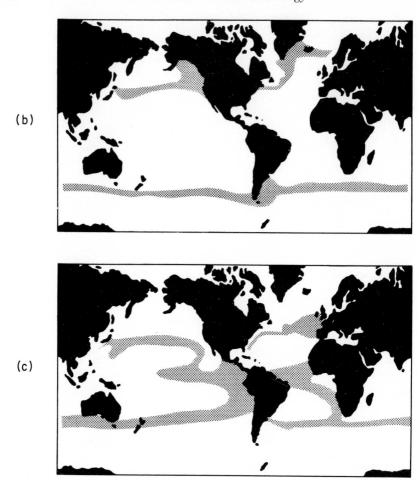

(b)

(c)

Fig. 4.4. (a) Present range of the euphausiid *Thysanoessa gregaria* compared with hypothetical circumstances of oceanwide warming (b) and cooling (c), both by 2·5°C. The 7°C and 11°C isotherms at 200 m depth, associated with present limits of distribution of *T. gregaria*, are considered limiting for the purpose of the extrapolation. With warming (b), the 7°C isotherm (at the high-latitude boundary) moves to position of present 4·5°C isotherm, and the 11°C isotherm (at the low latitude boundary) moves to position of present 8·5°C isotherm. With cooling (c) the limiting 7°C and 11°C isotherms move to positions of present 9·5°C and 13·5°C isotherms, respectively (after M. W. Johnson and E. Brinton[152]).

abundant in the colder conditions. Bradshaw also observes that the laboratory studies indicating that particular species thrive best under particular environmental conditions are reflected in the oceans themselves.[30] Figures 4.2 and 4.3 show that the distribution of planktonic Foraminfera in the North Pacific Ocean follows closely the distribution of water masses, defined themselves by particular physical properties, such as ranges of temperature and salinity.

The euphausiid *Thysanoessa gregaria* is at present confined to two mid-latitude belts in the oceans and one puzzle is that identical planktonic species should be isolated north and south of the equator.[152] If there had been time, independent selective adaptations might have led to distinctive populations. It has been suggested that the distribution would be radically altered if the near-surface temperatures changed by a few degrees. If the temperature of the oceans has been generally lower the now-isolated groups of this euphausiid would form a single population as shown in Fig. 4.4.

An empirical approach has more, though not complete success; here an attempt is made to relate the plankton observed living in the sea with the skeletal remains in the surface sediment beneath. (To relate, that is, the biocoenoses to the thanatocoenoses.) Among a number of studies of this nature is one in which a close relationship was established between the diatom biocoenoses and thanatocoenoses in the western part of the Bering Sea.[153] There is correspondence between the living assemblages and the bottom skeletal assemblages of Foraminifera in the equatorial part of the Atlantic Ocean, provided (as with the diatoms) that some allowance is made for dissolution of the tests after the death of the organism.[239]

Phleger has discussed the problems of foraminiferal ecology,[223-4] and points out that quite the wrong conclusions may be drawn innocently from apparently unambiguous evidence. Off river mouths, for example, the distribution of large living foraminiferal populations coincides with very rapid detrital deposition, so that large "standing crops" of Foramini-fera are diluted by sediments after burial. Consequently the record pre-served in rocks which are the end product, appears sparse in fauna, and the living conditions will be thought to have been marginal.

A second empirical approach is to try to relate the present distribution of skeletal remains of planktonic organisms at the surface of sediments to the distribution of water masses, and such a relationship is found among some Foraminifera in the North Atlantic Ocean;[223] they show a dependence upon the temperature of the water in which they live, and this dependence becomes significant when one is interested in the sequence of planktonic faunas (as represented now by the skeletal remains) found in the sedimentary sequence of a sediment core. This is the second stage of the investigation: having established that a relation-ship between temperature and the type or abundance of Foraminifera exists, such a relationship may be used in palaeo-climatological or palaeo-oceanographic investigations.

Two methods have been used for estimating the temperature of the water in which the Foraminifera whose tests are now found in sediment cores lived.

We can observe Foraminifera characteristic of cold and warm waters when living, and determine their relative abundance within sediment samples from a core. This allows us to say the water was colder or warmer than now. The direction of coiling of the tests of *Globigerina pachyderma* may also be used—the direction is temperature dependent.[72] Of course, if we are concerned with a stratigraphic boundary, a point in time, then this may not be seen in waters of high latitudes by directions of coiling, because any temperature change may have been too slight.

The second method depends upon determining the ratio of two of the isotopes of oxygen, O^{18} and O^{16} in the calcium carbonate of which the tests of the Foraminifera are made.[67,281] This was once thought to be an absolute method, and to give the temperature of the water in degrees. Doubt has been cast upon this recently, however, but it still remains of immense value, as will be seen.[342] In much of the rest of this book temperatures have been given in degrees as estimated by the oxygen isotope method, but this doubt must be borne in mind, and allowance made for it.

The distribution of the two isotopes between water and carbonate ions, in a solution which is precipitating calcium carbonate slowly, depends upon the temperature. The difference in the ratio O^{18}/O^{16} between the calcite and the water decreases with increase in the temperature at which it is precipitated.[67,281] If the isotopic composition of the water is known, or can reasonably be estimated, the actual temperature of precipitation of the calcite can be found. Emiliani has applied this technique most profitably. However, Shackleton has pointed out that the former isotopic composition of the ocean may not have been correctly calculated.[342]

The polar ice sheets are deficient in O^{18} by comparison with ocean water. This may be roughly ascribed to the extra mass of the O^{18} atoms — H_2O^{16} will evaporate from the ocean surface more readily than H_2O^{18}, and it is this evaporation which leads of course to precipitation subsequently. Consequently the growth and melting of ice sheets will change the isotopic composition of sea water. Shackleton, following E. Olausson, suggests that the whole record of isotopic changes recorded by planktonic Foraminifera may be due to changes in the isotopic composition of the oceans, caused by the waxing and waning of the continental ice sheets.

A curve of temperature against depth is shown in Fig. 4.5, where the

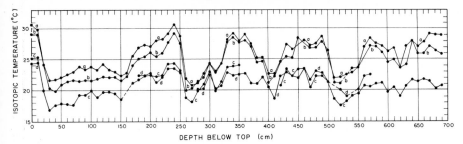

Fig. 4.5. Isotopic temperatures obtained from *Globigerinoides rubra* (a), *Globigerinoides sacculifera* (b), *Globigerina dubia* (c), and *Globorotalia menardii* (d). Numbers above the curves identify temperature units. Core A179-4 (16°36′N., 74°48′W., 2965 m) (after C. Emiliani [68]).

scale of degrees has been retained. Suppose the scale was not degrees, but the isotopic composition, from which the temperature is calculated. What now do the oscillations represent? They represent changes in the continental ice sheets—where the "temperature" is low the continental ice sheets were at a maximum, and where high, at a minimum. So the record is immensely valuable, for it traces the glacial–interglacial record of the continents, from core to core, throughout the world's oceans.

The study of the ratio O^{18}/O^{16} in equilibrium systems has not been confined to attempts to estimate temperature, and another problem in which measurements of O^{18} proved useful is that of the origin of dolomites. Dolomite, $CaMg(CO_3)_2$, is common in ancient sediments, but relatively less common in recent marine sediments. It could form in two ways—by addition of magnesium to calcite or aragonite already present, or by precipitation as the double carbonate. Experiments by Degens and Epstein [320] suggested that at normal water temperatures there should be a difference in the ratio O^{18}/O^{16} between coexisting dolomite and calcite. Studies of dolomite-calcite pairs actually found in sediments showed the difference to be much smaller than predicted— that is, the O^{18}/O^{16} ratio in dolomite is similar to that of the coexisting calcite. Consequently the dolomite must be formed in the first way mentioned—by addition of magnesium to calcite present already.

The use of the fossil representatives of the planktonic groups mentioned so far in stratigraphy of deep-sea sediments, depends upon the same factors that determine a fossil's (or fossil assemblage's) use in the stratigraphy of rocks found on the continental land masses. They must be abundant, well preserved, widespread and must have evolved rapidly. Foraminifera and coccolithophorids satisfy the first three requirements, at least at depths of water less than approximately 4 km, but the short time interval represented by sediment cores only some tens of metres long means that evolutionary changes which involve the appearance and disappearance of recognizably different species have been of lesser value so far to stratigraphy of deep-sea cores than has been the record of changes of temperature. Evolutionary changes which have been important include among them, for example, the disappearance of the discoasters, among the coccoliths, at the end of the Tertiary.[80] The Radiolaria have been used successfully in stratigraphic studies of Antarctic cores, in conjunction with studies of their magnetic properties.[340] This is discussed later in the chapter.

DETERMINATION OF THE ABSOLUTE AGE

Principles

The isotopes of a number of elements are radioactive and decay into products which may be stable, or may in their turn be radioactive.[160] For example, one isotope of carbon, C^{14}, decays into the stable nitrogen isotope, N^{14}, and such a simple decay scheme may be described in the following way.

Write $N_1^0 =$ number of atoms of the parent present in the system at time $t = 0$,

$N_1 =$ number of atoms of the parent present in the system at time $t = t$,

$N_2^0 =$ number of atoms of the daughter present in the system at time $t = 0$,

$N_2 =$ number of atoms of the daughter present in the system at time $t = t$.

Suppose that at time $t = 0$ there are no atoms of the daughter present, but only atoms of the parent isotope, that is

$$N_2^0 = 0,$$

The decay of the parent may be written

$$\frac{dN_1}{dt} = -\lambda N_1,$$

where λ is a constant, the decay constant. Consequently

$$N_1 = N_1^0 e^{-\lambda t}$$

and because, if the daughter is stable and does not decay further,

$$N_2 = N_1^0 - N_1$$

then

$$N_2 = N_1(e^{\lambda t} - 1).$$

The value of the decay constant λ can be measured, and we have available to us two ways of finding the value of t, the age of isolation of the parent in the system. The value of N_1 can be measured and the value for N_1^0 assumed, or both N_1 and N_2 can be measured. The first way is that of the method for age determination which uses the decay of C^{14}, the second is that of the method based upon the decay of Rb^{87} to Sr^{87}.

Such a system as this is rather simple; for example, the daughter atoms may be radioactive and several generations necessary before a stable end product is formed. The usefulness of any radioactive decay scheme in absolute dating depends upon a number of factors, of which two are of great importance: the rate of decay and the abundances of the radioactive nuclides. The decay must be substantial in times of the same order of magnitude as those we wish to measure, and we can express this quantitatively. The time taken for half the original atoms of the parent to decay is given by $T_{1/2}$, where

$$N_1 = N_1^0 e^{-\lambda T1/2}$$

and

$$N_1/N_1^0 = \tfrac{1}{2}.$$

Hence

$$T_{1/2} = (\log_e 2)/\lambda,$$

or

$$T_{1/2} = 0\cdot693/\lambda.$$

The abundance is obviously important. One potentially useful method could be based upon change in the ratio Al^{26}/Be^{10}, but although aluminium as Al^{27} is abundant, Al^{26} is not.[241]

Nuclides Generated by Cosmic Rays: C^{14}

A number of radioactive nuclides are generated in the atmosphere of the earth by cosmic rays, and among these are Be^{10}, C^{14}, Al^{26} and Si^{32}. Of these C^{14} has become a standard tool in geological and archaeological dating of material not older than about 70,000 years, for its half-life is 5570 years and it is produced in reasonable abundance. The C^{14} produced in the atmosphere is incorporated into living organisms as carbon dioxide and when the organisms die any exchange of the carbon of the organisms with their surroundings ceases. The C^{14} in a dead organism under study can be measured; if the C^{14} content of the atmosphere at present is known, and if the assumption is made that this content has remained constant during the length of time we wish to estimate, then the age of the dead organism which contains the C^{14} can be calculated. The application to problems in marine geology arises because the tests of many organisms (such as Foraminifera) are made largely of calcium carbonate. The assumption of constant rate of production of C^{14} can be investigated by using samples of known historical age, and by using a second method with the same material, such as Pa^{231}/Th^{230} (see below). This latter method is not quite equivalent, for it is applied to the inorganic fraction of a sample — such as the clay minerals, but C^{14} is estimated from the organically derived $CaCO_3$, and there may be an age difference between the two fractions. Sediment may be reworked or redeposited by water currents, to a greater extent for fine material — fine carbonate or clay particles — than for coarse; the age measured may be greater for the fine material than for coarse. The rate of accumulation of pelagic foraminiferal oozes may be only a few centimetres every thousand years, and reworking due to burrowing animals[7] has been estimated to be effective over distances of several centimetres. The resolution of C^{14} dating is thereby reduced, and it is in any case often of value only in the upper 50 or 100 cm of deep-sea sediment because of the low rate of sedimentation and the short half-life of C^{14}.

Other nuclides produced by cosmic rays may be useful; the half-life of Be^{10} is 2·5 million years, and that of Al^{26} is 740,000 years: Be^{10} by itself, or with aluminium as the ratio Al^{26}/Be^{10}, has a convenient half-life for many problems which involve deep-sea sediments, if the abundance problem could be overcome.[5]

The Decay of K^{40}–Ar^{40}

The decay of K^{40} to Ar^{40} and Ca^{40},

is useful in dating igneous rocks, even those with small potassium content such as basic rocks[10,110] and in the solution of geological problems, such as the origin of potassium bearing clay minerals in oceanic sedi-

Table 4.1 *The Decay Scheme of* U^{238}

The Uranium Family

Radioelement	Nuclide	Radiation	Half-life
Uranium I ↓	U^{238}	α	$4\cdot50 \times 10^9$ y
Uranium XI ↓	Th^{234}	$\beta-$	$24\cdot10$ d
Uranium X2 ↓	Pa^{234}	$\beta-$	$1\cdot14$ m
Uranium II ↓	U^{234}	α	$2\cdot52 \times 10^5$ y
Ionium ↓	Th^{230}	α	$7\cdot52 \times 10^4$ y
Radium ↓	Ra^{226}	α	1622 y
Raemanation, Radon, Niton ↓	Rn^{222}	α	$3\cdot825$ d
Radium A	Po^{218}	$\alpha, \beta-$	$3\cdot05$ m
99·96% ↓ 0·04% α			$26\cdot8$ m
Radium B	Pb^{214}	$\beta-$	$26\cdot8$ m
$\beta-$ Astatine	At^{218}	$\alpha, \beta-$	$1\cdot5$ s–2 s
99·99% 0·01% $\beta-$ α Radon	Rn^{218}	α	$0\cdot019$ s
Radium C	Bi^{214}	$\beta-, \alpha$	$19\cdot7$ m
0·04% 99·96% $\beta-$ α Radium C′	Po^{214}	α	$1\cdot637 \times 10^{-4}$ s
Radium C″	Tl^{210}	$\beta-$	$1\cdot32$ m
Radium D	Pb^{210}	$\beta-$	25 y
Radium E	Bi^{210}	$\beta-, \alpha$	$5\cdot02$ d
~100% ~5 × 10⁻⁵% $\beta-$ α Radium F Thallium	Po^{210} Tl^{206}	α $\beta-$	$138\cdot3$ d $4\cdot23$ m
Radium G	Pb^{206}	Stable	—

ments.[52,146,171] Applied to cores of sediment from the ocean basins, the precise "age" to be assigned to the mineral may not be desired, but we may want to know if the clay mineral is rather "young" (say 0–10 million years old) or rather "old" (say 300–400 million years old). The distinction may answer questions which concern the origin of the clay minerals, or the stability of the potassium ions in their sites within the mineral lattice.

The age of the mineral can be calculated if the K^{40} content and the radiogenic Ar^{40} content are known, and if the two decay constants for the dual decay scheme are known. A number of conditions must be satisfied for the age assigned to be reliable. There must have been no Ar^{40} trapped in the mineral at the time of its formation, and no addition or subtraction of the K^{40} or Ar^{40}, except through radioactive decay, since the time of formation. The half-life of K^{40} is about $1\cdot3 \times 10^9$ years and the method has been used for ages as young as a few hundred thousand years.[75]

A method rather similar in principle is that which uses the decay scheme,

$$Rb^{87} \rightarrow Sr^{87}.$$

The half-life of Rb^{87} is about 5×10^{10} years,[234] and one use of this decay scheme will be seen in Chapter 7.

The Decay of Isotopes of Uranium

The decay schemes of the two uranium series U^{238} and U^{235} are presented in Tables 4.1 and 4.2. The nuclides generated in this decay have proved useful in the determination of the ages of sediments on the floor of the deep-sea, in the range of time up to a few hundred thousand years. One limitation of the C^{14} method is the small length of time over which it is useful, and hence the short length of sediment core to which it can be applied. (The maximum age of 70,000 years, mentioned above, is an extreme limit; the maximum ages found commonly in the literature are approximately 30,000 years.) The methods which depend upon the nuclides generated from U^{235} and U^{238} are more useful in so far as they can be applied to samples of greater age.

The parts of the decay schemes relevant here are these:

(1) $U^{238} \longrightarrow Th^{234} \longrightarrow Pa^{234} \longrightarrow U^{234} \longrightarrow Th^{230} \longrightarrow Ra^{226}$
(2) $U^{235} \longrightarrow Th^{231} \longrightarrow Pa^{231} \longrightarrow Ac^{227} \longrightarrow Th^{227}$

Examination of Tables 4.1 and 4.2 will show that the half-lives of these

Table 4.2 *The Decay Scheme of* U^{235}

The Actinium Family

Radioelement	Nuclide	Radiation	Half-life
Actinouranium	U^{235}	α	$7 \cdot 07 \times 10^8$ y
Uranium Y	Th^{231}	$\beta-$	$25 \cdot 6$ h
Protactinium	Pa^{231}	α	$34 \cdot 3 \times 10^3$ y
Actinium	Ac^{227}	$\beta-, \alpha$	$27 \cdot 7$ y

98·8% | 1·2%
$\beta-$

Radioelement	Nuclide	Radiation	Half-life
Radioactinium α	Th^{227}	α	$18 \cdot 6$ d
Actinium K	Fr^{223}	$\beta-$	21 m
Actinium X	Pa^{223}	α	$11 \cdot 2$ d
Actinon	Rn^{219}	α	$3 \cdot 92$ s
Actinium A	Po^{215}	$\alpha, \beta-$	$1 \cdot 83 \times 10^{-3}$ s

~100% | ~5×10^{-7}%
α

Radioelement	Nuclide	Radiation	Half-life
Actinium B $\beta-$	Pb^{211}	$\beta-$	$36 \cdot 1$ m
Astatine	At^{215}	α	10^{-4} s
Actinium C	Bi^{211}	$\alpha, \beta-$	$2 \cdot 16$ m

$\beta-$

Radioelement	Nuclide	Radiation	Half-life
Actinium C' α	Po^{211}	α	$0 \cdot 52$ s
Actinium C''	Tl^{207}	$\beta-$	$4 \cdot 76$ m
Actinium D	Pb^{207}	Stable	—

nuclides fall into three groups: those very long (U^{238}, U^{235}), those of the order of tens or hundreds of thousands of years (U^{234}, Th^{230}, Pa^{231}, Ra^{226}) and those very short, of the order of days or less.

The peculiarities of the distribution of these nuclides between sea water and sediment lead to the possibility of dating the sediments. A given concentration of U^{238} and U^{235} in sea water should lead to specific concentrations of Th^{230} (ionium) and Pa^{231} (protactinium) respectively in sea water, if radioactive equilibrium is maintained. It is found however that the concentrations of these latter two nuclides is much lower in sea water than would be suggested by the uranium isotopes' concentrations, and conversely their concentrations in the surface layers of sediments is much higher than can be accounted for by the concentration of uranium in the sediment.[160,225] This phenomenon leads to Ra^{226} concentration which is greater than its uranium-supported concentration could be. One method of dating a core of sediment might then be to measure the radium content or the ionium content of samples taken at intervals down the core. The success of such a method will depend however upon a number of factors, among which are the following. The rate of precipitation of the Th^{230} and its incorporation in the sediment must remain constant over the time interval being measured; this implies that the uranium content of sea water has also been constant. If Ra^{226} is used, then it must not migrate vertically within the sediment. Picciotto has pointed out[225] that these assumptions are unlikely or demonstrably false. These difficulties, and others, may be overcome if the ratio Pa^{231}/Th^{230} is used, for the two nuclides are likely to have similar chemical properties, and are both produced from the same element, uranium. Therefore, changes in the properties of the sediments and of the rate of supply of uranium to the oceans will not affect changes of the ratio with time caused by radioactive disintegration. This method has been described in some detail by Rosholt and others.[242]

A rather similar approach is to measure the ratio Th^{230}/Th^{232} (commonly written as Io/Th.[99,100] Th^{232} is naturally occurring thorium with a half-life of $1·41 \times 10^{10}$ years, which is so long that the isotope may be thought of as stable for times up to a few hundred thousand years. Both Th^{230} and Th^{232} should be removed simultaneously from the water. The principal assumptions which have to be made are:[100] (1) The Th^{230}/Th^{232} ratio in the water mass adjacent to the sediment in a given ocean basin has remained constant over the interval of time being measured. (2) There is no fractionation of the two thorium isotopes between sea water and the solid phases deposited which contain thorium. (3) The materials analysed do not contain detrital substances of continental or volcanic origin with significant contributions of ionium or thorium. (4) There is no migration of thorium in the sediment.

In summary, three methods are now being used to estimate the time of deposition of sediment upon the sea floor, C^{14}, Th^{230}/Th^{232} and $Pa^{231}/Th.^{230}$ These methods overlap in range of application, and can be checked one against another. Some cross-checks show agreement,[242] some do not.[99,241,242] At the present time it is not possible to say whether the rate of sedimentation in particular regions of the North Atlantic is close to millimetres per thousand years or centimetres per thousand years.[101] The disagreement can in some cases be ascribed to particular causes. In the Th^{230}/Th^{232} method the contributions of the Th^{230} and Th^{232} precipitated from sea water are found by leaching the sediment, assuming that any Th^{232} in detrital grains derived from the continents is not affected by the leaching process. If it is affected and enters the leached solution the Th^{232} contribution will be erroneously high, and the age deduced too great. This will lead naturally to an apparently low rate of deposition.

MAGNETIZATION OF SEDIMENTS: MAGNETIC STRATIGRAPHY

Rocks can acquire a magnetization in the direction of the ambient earth's field in a number of ways. Lavas take on a thermoremanent magnetization when their constituent magnetic minerals, such as magnetite, cool to temperatures below their Curie points. Some sediments are magnetized when deposited. Magnetic particles falling through water are oriented in the direction of the earth's field. Others may be magnetized chemically—as a magnetic mineral is deposited out of solution it is magnetized in the direction of the earth's field.

The earth's magnetic field changes polarity at intervals of time of the order of one million years.[51,148,339,340] Consequently, if sediments on the ocean floor are magnetized at or soon after the time of their de-. position they should preserve a record of the changes in polarity. The times of changes in polarity can be dated in one of two ways. First, the ages of lavas on the continents can be found by the K-Ar method (described earlier in this chapter), and if their directions of magnetization are also found a time-scale of reversals can be established. Second, a time-scale of reversals can be set up from the cores of sediments themselves, using one of the methods which involves ionium, for example. The time-scales of reversals found in these two ways agree well with each other.[339,340] Reversals of magnetization were first found in sediments by C. G. A. Harrison.

The reversals can now be used to date many sediment cores in the oceans in a precise way. The changes in polarity provide fixed reference points within each core.

Studies of the record of reversals preserved in sediments are important for several reasons. A palaeontological time-scale of reversals has been established using Radiolaria.[339] This was done for many cores, and the contemporaneity of the changes in polarity established from core to core. Such an experiment checks that sediments do acquire their magnetization at or close to the time of deposition.

Dr. R. Uffen has suggested that a mechanism for evolution is provided by reversals of the earth's magnetic field. During a reversal the earth might be exposed to more intense radiation than is normal. The extra radiation might lead to genetic changes. This theory can be tested from the evidence of the sediment cores. So far it has not been proved.

The finding of reversals in cores of sediment checked independently the time-scale of reversals established about ocean-floor spreadings, considered later in the book (see Chapter 8 and 9).

MINERALOGY AND GEOCHEMISTRY OF PELAGIC SEDIMENTS

The chemistry of the system air–sea–rock should ideally be considered as a whole. If the chemical composition of the system is known and the constants of the (many) possible reactions are known then it might be possible to deduce the distribution of the elements between the parts of our system and deduce the mineralogy of the rocks, mainly sediments, at the bottom of the sea.[253] There are gaps in our knowledge which make such an ideal calculation only approximate, but it is well to remember that sediments (and other rocks) do not exist in isolation: their constituents are bounded above by sea water and air. The minerals which make up sediments are derived from five sources:

(1) from the continents, and are called *terrigeneous*;

(2) from sea water, by crystallization (and are called by Arrhenius *halmeic*);

(3) from volcanic eruptions, in and outside the sea, and are called *pyroclastic*;

(4) from living organisms, and are called variously *biotic* or *biogenic*; and

(5) from extra-terrestrial sources, and are called *cosmic*.

These components are either in equilibrium with the surrounding sediment, the interstitial water between the sediment particles or sea water, or will, if there is time and opportunity, attain equilibrium in the course of appropriate chemical reactions. For example, calcium carbonate as calcite is in equilibrium with the ocean down to certain depths at particular temperatures, but dissolves at greater depths and lower temperatures. The relative contribution to the composition of any particular sample of sediment which will be made by these five sources will vary, although the contribution from cosmic sources is likely to be always small. It is more or less obvious that sediments near continents will contain a higher proportion of terrigeneous material than those sediments which are farther away; for example, the percentage of quartz grains derived from the Sahara desert found in the sediments of the

Cape Verde basin decreases with distance from the shore. However, it is not so obvious why the latitudinal dependence of quartz concentrations in the surface sediments of the Pacific Ocean should correspond to that of the arid areas of the world, and also be greatest in deposits furthest from land.[238,323] In this case the distribution of ocean currents cannot be invoked to explain the phenomenon; Rex and Goldberg suggest that the peculiarities of the distribution are caused by atmospheric transportation by tropospheric winds, which are also latitudinally zonal (Fig. 4.6).

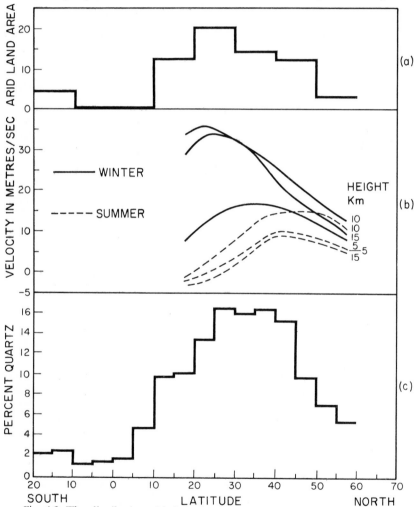

Fig. 4.6. The distribution with latitude of the arid areas of the world (a), the velocities of high altitude winds (b), and the per cent of quartz in east Pacific sediments (c) (after R. W. Rex and E. D. Golberg [238]).

Such environmental influences are of great importance to the composition and distribution of marine sediments. The trenches which surround parts of the Pacific Ocean prevent continental detritus from reaching the deep-sea floor, to form abyssal plains, to the same extent that it does in the Atlantic Ocean (where trenches at the continental margin are less numerous). The smaller size of the Atlantic Ocean, coupled with the greater discharge of rivers from the continents and the ease of access to it of glacially transported sediment, has led to greater rates of deposition in this ocean than in the Pacific Ocean.

The only elements that occur free in pelagic sediments are nickel and iron together, in the form of droplet-like rounded particles, which because of their shape and chemical composition have been ascribed to an extra-terrestrial origin.[145] The spherules are partly oxidized and are associated with chondrules of crystalline olivine and pyroxene; the nickle–iron spherules and the silicate chondrules represent probably the material released on disintegration of meteorites, rather than "cosmic-dust".[7] Their identification as extra-terrestrial material is complicated, because of other sources of "black spherules", such as blast furnaces and volcanoes.[89]

The spatial distribution of carbonates in sediment, laterally and vertically, is affected by the solution under high water pressures and at low temperatures of the principal carbonates calcite and aragonite. The carbonate content of sediment depends therefore upon the rate of production in the water—the productivity of the water with respect to calcareous organisms such as Foraminifera and coccolithophorids, the depth and the temperature of the water in which the skeletal remains are deposited, and the dilution of the carbonates by other minerals such as clays. If the calcareous skeletal remains are exposed for a long period of time it is more likely they will dissolve, and one consequence of this is that at great depths in the Atlantic Ocean the calcium carbonate content of sediment is greater than in the Pacific Ocean (Fig. 4.7). Turekian suggests that at depths in the Atlantic Ocean below 4000 m calcium carbonate will in general dissolve because the water is unsaturated, but that the variety of particular local conditions which are found lead to a great range of values of calcium carbonate percentage at depths over 4000 m.[277]

The distribution of the rock-forming silicates within the sediments of the ocean basins depends upon the relative contributions from continental and oceanic sources, the physical and chemical conditions of the water and upon the minerals themselves. The problem to be solved may often be to decide whether or not a mineral was formed *in situ*, or if not, where did it come from and how was it transported? Igneous rocks

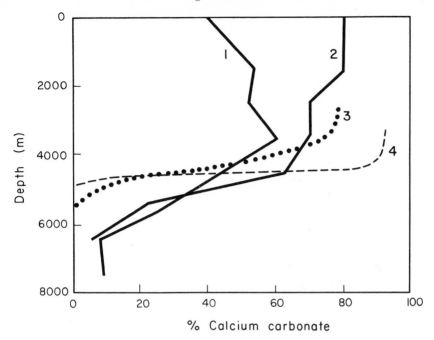

Fig. 4.7. Calcium carbonate content of the tops of sediment cores. The average values are shown for all deep Atlantic (1), mid-Atlantic (2), all Pacific (3) and mid-equatorial Pacific (15°N.–5°S., 120°W.–135°W.) (4). Curves from Turekian[277] and Bramlette.[31]

weather to a variety of minerals, and the minerals formed depend upon the environment. Some minerals are relatively stable and are not altered — rutile and pyrophyllite for example. Quartz being relatively stable and common on continents, is ubiquitous in pelagic sediments, wind-borne or water-borne. The end products of the weathering of the less stable minerals, the feldspars for example, include clay minerals among them. The particular clay mineral produced varies with the parent rock and the environment. Montmorillonite is a product of the weathering of basic igneous rocks in the sea, and it is found in sediments in high concentrations near volcanic regions. Kaolinite is the commonest clay mineral, and is derived from the weathering of feldspars — hence from igneous rocks. Illites are the dominant clay minerals of shales and mudstones, and break down to montmorillonite in water. Gibbsite ($Al(OH)_3$) arises from the weathering of basic igneous rocks in the tropics (it is a constituent of the aluminium oxide deposits called bauxite, found, for example, in Jamaica).[58] Some details of the weathering of volcanic rocks in sea water are given in Chapter 7. Clay minerals could therefore arise by weathering of igneous rocks on land and subsequent transport as clay minerals to the sea bottom, by weathering of igneous rock on the sea bottom, or by generation *in situ*, and we have to find out in any particular set of circumstances which of these possibilities is correct. Minerals which are resistant, rutile, the garnets and pyrophyllite for example, could be useful as tracers if their source areas could be clearly defined.[22]

The distribution of alumino–silicates such as the clays illite and mont-morillonite, and the mica-like mineral chlorite, depends then upon the authigenic contribution and the continental contribution.[98] Clay minerals transported by water from the continents are found near the continental margins of the North Pacific Ocean, whereas in the central part of this ocean the clay mineral illite is associated with quartz transported there by wind; the association suggests illite was transported in this way too.[103] The rate of sedimentation in the South Pacific Ocean is lower than in the North Pacific because the area of adjacent continental land mass is smaller, and this, if only through lower dilution, leads to an increase in the concentration of authigenic minerals such as the zeolite phillipsite, and also montmorillonite formed by alteration of volcanic products.

Another approach to the problem of origin of clay minerals is to determine the K^{40}–Ar^{40} "ages" of clays from surface sediment samples.[146] It is known that the surface sediment of the ocean floors was deposited relatively recently, but the K^{40}–Ar^{40} "ages" of the clay minerals are found to be in the range 200–400 million years, which shows that they must be detrital minerals — brought from the continents as clay minerals, or from some part of the ocean floor sufficiently old. The importance of both the conditions on the continent and the oceanic circulation upon the present distribution of clay minerals in surface samples on the sea floor has been shown by Biscaye.[22] He found that the distribution of the minerals kaolinite and gibbsite, both end-products of the weathering of igneous rocks, and gibbsite typical of the tropics,[99] depends upon the local supply from the continents, and that the distribution of gibbsite was modified by the Gulf Stream. This is an example of a neglected study — the influence of ocean currents as transporting agents, rather than as discrete water masses in which a particular flora or fauna will flourish. Geological studies tend to emphasize temperature and depth of water when considering fauna, and go on to assume that temperature bears a simple relationship to latitude, which is not necessarily true.

The influence of ocean currents in the transport of sediments and the shaping of submarine topographic features is seen in studies of the continental margin of eastern North America,[328] considered in Chapter 8, and of Baffin Bay.[330] The study by Heezen and others of the continental margin emphasizes the effect of surface and deep ocean currents as agents of transportation of sedimentary detritus within the sea itself. Marlowe's study of Baffin Bay, and other studies by the Geological Survey of Canada in Hudson Bay, shows the effect of surface currents upon ice and the debris carried by it—the ice in the form of icebergs for example. The surface water circulation in Baffin Bay (Fig. 3.2) is counterclockwise, the relatively warm west Greenland current flowing northwards to join water from the Arctic Ocean flowing south along the east side of Baffin Island. Icebergs which calve off west Greenland or in the sounds which lead into Baffin Bay follow these currents, and the sediments on the floor of the bay, formed in part by unloading of the rock debris from the icebergs, reflect the source areas from which they calved. Fragments of limestone characteristic of the land areas in the northern and western parts of Baffin Bay are found in the surface sediment far to the south of the source areas on the western side of the bay. Using such fragments as tracers, Marlowe showed that at one time the water circulation must have been different; the limestone fragments of the deeper and therefore older parts of sediment cores are found further to the east than are those in the surface sediment. The water circulation in Hudson Bay is also counterclockwise and the distribution of sediments there is controlled by its effect upon the ice movement. In both situations, apart from the intrinsic interest, one lesson to be learned is that deduction from data gathered from sediments on continents relating to past oceanographic conditions is a difficult exercise.

APPLICATIONS OF STUDIES OF PELAGIC SEDIMENTS

Studies of pelagic sediments concerned with stratigraphy or with past climates, or with the faunal changes in the sediment, are related one to another.[33] The major problem in such studies is the dynamic rather than static characteristics of pelagic sedimentation, as has been forcefully pointed out by Heezen;[114] a suite of sediment cores is needed which have accumulated particle by particle throughout the whole time represented by the core, not interrupted by slumps, or affected by winnowing of fine material by currents, and such cores are not common. The author made calcium carbonate analyses of three sediment cores collected near the foot of seamounts which rise to the west of the Iberia abyssal plain in the eastern basin of the North Atlantic Ocean. They were taken as close together as was possible by conventional navigation, yet showed very little correlation in lithology (as distinguished by the carbonate analyses). One disturbing influence will be bottom currents; there is evidence from photographs of the sea floor that water currents with velocities in the range of centimetres per second not only exist, but are systematically related to the water masses above.[124] Currents of this velocity can transport particles along the sea bottom, and modify the depositional history of wide areas. In deep-sea cores well-sorted foraminiferal sands may be found and they will sometimes owe their origin to winnowing.

The account just presented of the distribution of some clay minerals refers to minerals not, largely, formed *in situ*, but land-derived. A study of the sulphur compounds in the basin sediments off southern California showed by contrast, that minerals such as pyrite can be authigenic.[155] Insufficient sulphur is present as sulphate in the interstitial waters of the sediment to account for the total sulphur content, and it is proposed that sulphur is extracted out of the overlying water at the sediment-water interface by biological sulphate reduction.

Many studies have shown that the sequence of faunas in a sediment core not affected too much by dissolution of the skeletal remains is one of alternation of cold- and warm-water assemblages; these may correspond to high- and low-latitude assemblages.[73, 223] The pattern of these alternations has been used to correlate cores of sediment widely distributed throughout the oceans, to correlate in terms of time, that is, which assumes that all the changes in climate which led to the changes in faunal content are represented in the cores and that the sampling within the cores in the course of the study was sufficiently closely spaced. It assumes too that the climatological changes and faunal changes are isochronous throughout. An example of correlations based upon faunal changes is shown in Fig. 4.8, from studies by Ericson and Wollin.[74] Climatological changes are seen too in the oxygen isotope experiments designed to find the temperature at which the calcium carbonate tests of Foraminifera were deposited, and a series of results from one core is shown in Fig. 4.5.

The times at which these faunal changes took place, and at which the water temperatures changed, may be found by radioactive dating: the two principal methods, C^{14} and Pa^{231}/Th^{230} are applied to different components of the sediment, the biogenic components and the inorganic components respectively, and these may have had histories after deposition different one from another. The inorganic fraction, being relatively fine-grained, may be reworked and transported by water currents more easily than foraminiferal tests, which are relatively large. A combination of all these techniques has led then to information concerning past changes in climate, the relative rates of deposition of sediments in different parts of the ocean basins, and the stratigraphy of one sedimentary section compared to another.

The apparent amplitude of the temperature fluctuations of the surface

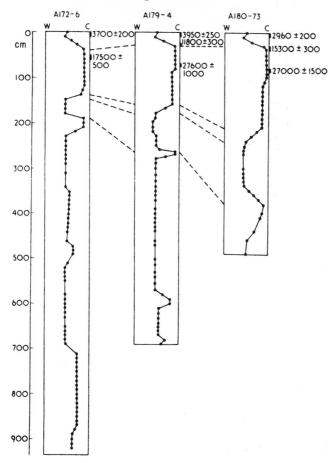

Fig. 4.8. Temperature variations and correlation of faunas in two cores from the Caribbean, A172-6 and A179-4, and one core from the equatorial Atlantic. The curves are based on the proportion of the number of warm-water to cold-water planktonic Foraminifera. W indicates relatively warm climate and C relatively cold climate. The blacked-out zones indicate the parts of the cores which were sampled for radiocarbon age determinations, and the numbers to the right of these zones are radiocarbon dates in years by H. E. Suess. The broken lines connect faunal and climatic changes which are considered to have taken place at the same time (after D. B. Ericson and G. Wollin [74]).

water during the glacial ages of the Pleistocene is different in different oceans (but see page 60). In the Pacific Ocean it is about 4°C, in the Atlantic Ocean about 6 or 7°C and in the Mediterranean Sea about 12°C.[68] We have seen[152] that the distribution of certain euphausiids in the Pacific Ocean can be explained if the temperature of the water in which the organisms lived was at some time lower than it is now, and it may be that similar studies in the Atlantic Ocean would yield distributions of organisms which also requires temperature changes to explain their present pattern. One might expect the effects to be more accentuated in the Atlantic Ocean, because of the larger temperature fluctuations. The change in temperature of the bottom water has been estimated from oxygen isotope analyses of benthonic Forminifera.[68] Whilst studies comparable in scope to those which have been made on planktonic Foraminifera have not been made, the studies suggest that the temperature of the bottom water of the Pacific Ocean has decreased by about 8°C since the Oligocene; this implies that an overall decrease in temperature of this magnitude occurred during the Tertiary. Such a change agrees with the change inferred from fossil floras and faunas of the continents.[67]

The biogenic components of the sediment found on the sea floor reflect the productivity of the overlying water masses. This is seen conspicuously beneath the Equatorial Divergence of the Pacific Ocean,[7] and beneath areas of high productivity in the Atlantic Ocean.[277] The high carbonate content of the sediments beneath the Divergence decreases to the north, away from the Divergence, and this is reflected in a number of parameters which indicate productivity (Fig. 4.9). For

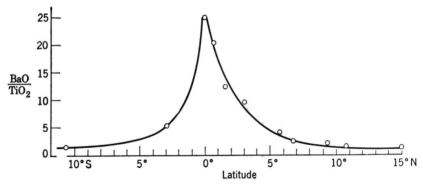

Fig. 4.9. The change in the ratio BaO/TiO₂ in Postglacial sediments with latitude near the equator in the Pacific Ocean, along 130°W. longitude approximately (after E. D. Goldberg and G. Arrhenius[98]).

example, associated with high productivity, defined by high diatom content of the sediment, is a relatively high content of barium, and the barium content of sediments decreases to north and south of the equator; this is shown in Fig. 4.9. Moreover, the temperatures of the water in which the planktonic Foraminifera live increases away from the Divergence (at which cold, nutrient-bearing water is brought up to the surface), and this is true not only for those of the present day, but for Pleistocene Foraminifera during glacial ages. A sequence of sediment cores some hundreds of metres long only ("only" with respect to longer drill-holes which have been suggested) would establish the position of the Equatorial Divergence at various times past—in the Tertiary, for example, and hence establish the past positions of the equator and the geographical poles.[7]

CHAPTER 5

Abyssal Plain Sediments

INTRODUCTION

Sediments deposited on abyssal plains differ in a marked way from those which are deposited on sites topographically higher than the plain, such as seamounts and abyssal hills. Their main characteristics are illustrated by those of the Iberia abyssal plain, described in the following section.

The abyssal plains merge gently with the rises at the feet of prominent topographic features where there is a supply of sediment which can be dislodged to seek a lower-level. Such source areas include among them the continental margins or the sea floor which surrounds a trench. Among the abyssal plains are, for example, the Iberia and Biscay abyssal plains at the foot of the continental margin of western Europe and that in the bottom of the Puerto Rico trench. The hypothesis that has been put forward to account for the characteristic topography of the plains and of the fans and cones of sediment which merge into them in some places, is that of the transport of sediment to the plains by turbidity currents.[9, 115, 151] These are currents of sediment-laden water which flow down-slope because their density is higher than that of the water above. Their density is not known precisely, but Menard suggests that the density contrast with water is rather small—perhaps $0 \cdot 1$ g/cm^3 or less, and they are cloud-like rather than oozing-liquid like.[199] In motion they must be in the range of a few metres to a few tens of metres thick and they flow with velocities of the order of kilometres per hour; the exact thicknesses and velocities depend upon interpretation of imperfect data.[117-19, 199] The sediment may be provided by a number of sources initially—river borne, or transported by long-shore currents; a good example of a supply available over long periods of time is Sable Island off Nova Scotia (Figs. 3.6, 3.7 and 3.8). This is situated on the outer edge of the continental shelf, a few kilometres from the present shelf-break and is the only emergent point on the outer banks of the Nova Scotian Shelf. It is some 40 km long and consists entirely of unconsolidated sands which are Pleistocene in age and of glacial and wind-blown origin

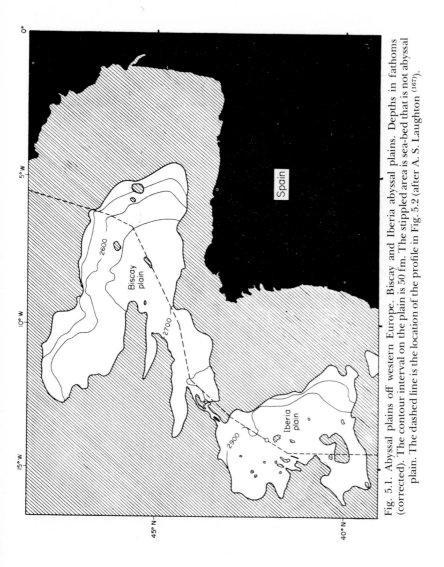

Fig. 5.1. Abyssal plains off western Europe. Biscay and Iberia abyssal plains. Depths in fathoms (corrected). The contour interval on the plain is 50 fm. The stippled area is sea-bed that is not abyssal plain. The dashed line is the location of the profile in Fig. 5.2 (after A. S. Laughton [167]).

which accumulated when sea-level was lower. The island is preserved
still, in spite of winds and sea, because of a peaty palaeosol underlying
the dunes.[197] To the north-east of the island is the "Gully",[188] a conspicuous indentation in the continental margin on even small-scale
charts, and Sable Island and Sable Island Bank provide a continuous
supply of sediment to fall down it. The turbidity-current can be initiated
in a number of ways, but slumping is probably common, and will arise
whenever the sediment's shear-strength is exceeded, whether or not a
positive trigger such as an earthquake is provided conveniently.[199] It is
not the slump which is the turbidity-current, of course, rather, the one
becomes the other; this is clear from the characteristics of the sediment
deposited at the foot of the topographic elevations.

The turbidity-current travels down-slope. As it slows down in the area
of deposition the load will be deposited, coarse material initially,
followed by increasingly fine-grained material as the velocity diminishes;
any one bed which is supposed to originate in this way is observed to
range up to about 1m in thickness. If the turbidity-current is travelling
at high speed, the hypothesis is that it will, in some circumstances, erode
the surface over which it travels.[115]

THE SEDIMENTS OF THE IBERIA ABYSSAL PLAIN

The Iberia abyssal plain lies west of the continental margin of Spain
and Portugal, in the eastern basin of the North Atlantic Ocean (Figs. 5.1,
5.2 and 5.4). Abyssal hills and seamounts rise to the west of the plain and

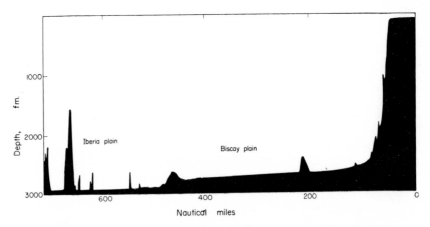

Fig. 5.2. Profile of Biscay and Iberia abyssal plains showing difference of levels.
Depths in fathoms (corrected). Vertical exaggeration is approximately 110:1.
For location of profile, see Fig. 5.1 (after A. S. Laughton [167]).

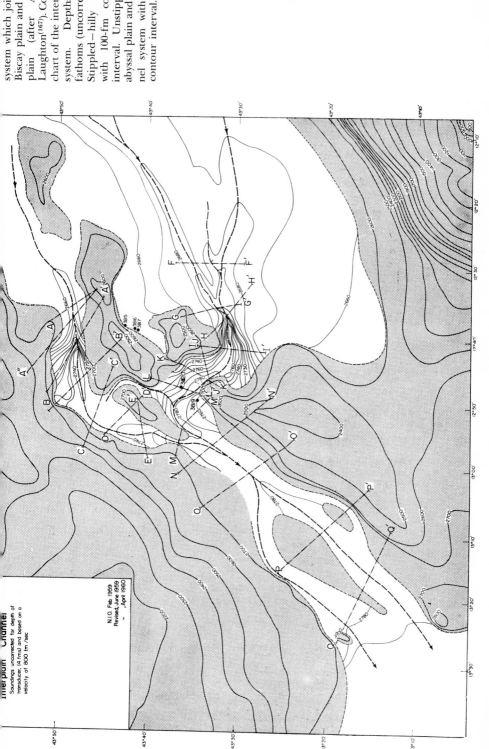

system which joins the Biscay plain and Iberia plain (after A. S. Laughton[167]). Contour chart of the inter-plain system. Depths in fathoms (uncorrected). Stippled—hilly area with 100-fm contour interval. Unstippled—abyssal plain and channel system with 5-fm contour interval.

Inter-plain Channel

Soundings uncorrected for depth of transducer, (4 f.ms) and based on a velocity of 800 fm/sec

N.I.O. Feb. 1959
Revised, June 1959
"   ,April 1960

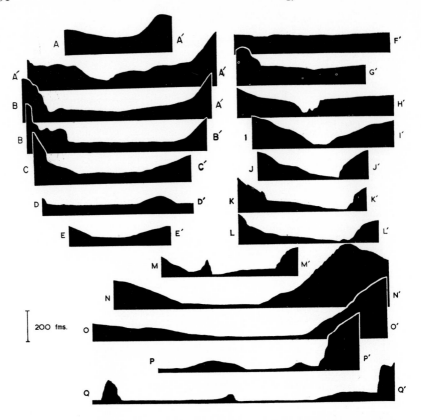

Fig. 5.3 (b). Theta Gap: an inter-plain channel system which joins the Biscay plain and Iberia plain (after A. S. Laughton[167]). Profiles of inter-plain channels. Locations of profiles indicated in Fig. 5.3 (a). Vertical exaggeration is 8·4:1.

within the plain itself; to the north it is connected by an abyssal gap or inter-plain channel to the Biscay plain[125] (Figs. 5.3 a and b).

Sediment cores taken from the hills and seamounts in the western part of the Iberia plain are very different from those taken from the plain itself[158] (see Figs. 5.4, 5.5 and 5.6). Those taken from the tops of the hills consist of lutites (very fine-grained sediments) alternating with calcareous oozes, and the larger rock particles within the sediment consist of fragments of volcanic rocks, quartz grains and some fragments of "granitic" types of rock. The sediment on the plain itself is made up of a rhythmic sequence of calcareous ooze or brown lutite, and a grey-black layer which is often graded with respect to grain size, the grain size of the constituent particles decreasing upwards. The base of each of the graded layers is sharp, marking distinctly the boundary

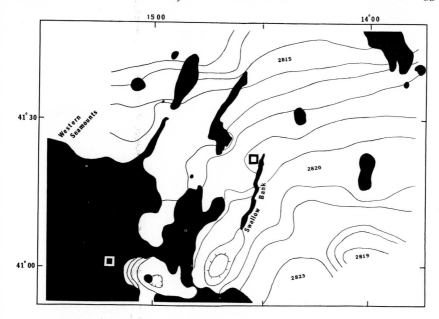

Fig. 5.4. The western part of the Iberia abyssal plain. The contours on the plain are in 1-fm intervals. Abyssal hills and mountains are shown solid black. The left-hand square is location of core 3738 of text and Fig. 5.5, the central square of core 3742 of text and Fig. 5.6. Depths in fathoms, uncorrected (assumed velocity 800 fm/s). Based on surveys by R.R.S. *Discovery* II (M. N. Hill, unpublished) (after a chart by A. S. Laughton[167]).

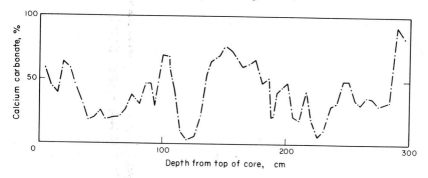

Fig. 5.5. Calcium carbonate analysis of a core from a seamount west of the Iberia abyssal plain. Note that this was not collected from the plain itself, and compare Fig. 5.6 (DyII station 3738, 41°00' N., 15°08' W., 2880 fm, corrected) (after M. J. Keen [158]).

between this layer and the one beneath. A typical cycle may be then:
(1) Calcareous ooze or a brown lutite: this may show evidence of burrowing, and have a mottled appearance. It merges gradually

with the layer beneath so that whilst the mottled sediment has at first a predominant white or brown colour, this changes until eventually it is predominantly grey or black, and finally there is no mottling to be seen at all.

(2) Grey or black graded layer: this may consist of organic remains, as calcareous tests or as lignite, of an assemblage of "continental" type rocks—granite and gneiss fragments, and abundant mica flakes. The grain size generally increases downwards. The base is sharp, by contrast with the top of the layer which merges into the calcareous ooze or brown lutite.

These features are shown diagramatically in Fig. 5.6. As in other forms of rhythmic sedimentation some of the parts of the cycle may be absent, and one grey layer may rest directly upon another, for example.

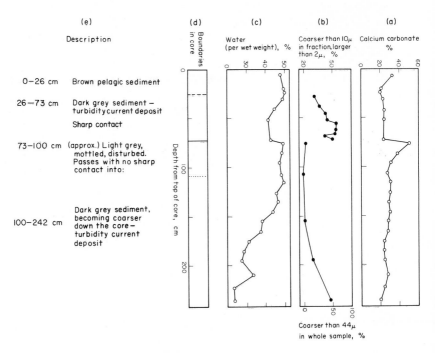

Fig. 5.6. Lithological features of a sediment core from the Iberia abyssal plain (after M. J. Keen [158]).

The hypothesis given in outline above suggests that successive turbidity-currents have crossed the plain bringing sediment from the continental shelves; normal pelagic sedimentation took place in between deposition from turbidity-currents. We would expect under these circumstances the bases of each deposit from the turbidity-current to be sharp and the tops to merge into the pelagic sediment deposited subsequently. We might look for two consequences of this hypothesis: (1) that the type of sediment may change systematically towards the source area, and (2) that the sediment deposition will gradually bury hills in the region of deposition. Both are found. The proportion of lignite increases in cores taken across the plain towards Portugal, and both seismic reflection studies and measurements of gravity at sea show that the thickness of sediments approaches 1 or 2 km beneath parts of some abyssal plains, thinning out at the edges of the depositional region; in this case this edge is at the foot of the Western Seamounts.

Such a mechanism leads to an explanation of a number of the features seen. The pelagic material in between each turbidity-current deposit will be dependent upon the prevailing conditions, and may be then a brown lutite or a calcareous ooze. The gradation between the pelagic sediment and the *lower* turbidity-current deposit will be smeared over by burrowing animals. The sharp bases of the turbidity-current deposits are not disturbed by burrowing.[73, 125]

The obvious source of sediments in the western part of the Iberia plain is the continental shelf of Spain and Portugal to the east. However, A. S. Laughton[167] noticed that at the northern end of the Iberia plain a gap exists through which sediment might be transported from the Biscay plain, and we will discuss this in the next section.

THE JUNCTION OF THE BISCAY AND IBERIA PLAINS

Laughton pointed out that although the Biscay and Iberia plains are separated by only 70 km there is a difference in height between them of 200 m. A survey from R.R.S. *Discovery II* showed that the two plains are connected by a channel (see Figs. 5.2 and 5.3) and that it is possible for sediment which originated on the continental shelves off Great Britain and France to have been transported into the Iberia plain through this channel. The features of the channel and the bottom contours of the two plains support this suggestion. Preliminary examination of the fauna of the turbidites from the Iberia plain also suggests that it may have originated in shallow water and in more northerly latitudes.

TURBIDITES OF FORMER TIMES

The abyssal plains represent basins filled in by sediment up to 1 or 2 km thick, a large part of which was apparently transported by turbidity-currents. The accumulation of sediments in trenches, such as the Puerto

Rico trench, will have been due in part to transport by turbidity-currents. However, in-filling of trenches is not necessarily down the slope of the walls of the trenches, for ridges and troughs parallel to the trench may catch such material on the walls at higher levels than the bottom. The Aleutian trench is filling up from one end.[199] "Fossil" abyssal plains have been mentioned by Ewing, Ewing and Talwani, underlying the present abyssal plains.[124] One significance of such a feature is that it was once nearly horizontal, so that deformation after its formation will be recorded clearly.

Characteristics of the sediments transported by turbidity-currents and deposited on the plains, which are important in the interpretation of the lithology of ancient sediments found on the continents, are the alternation of graded beds and pelagic sediments, and the presence of shallow-water fauna and flora in a deep-water environment. In addition, photographs of the deep-sea floor have shown that ripple-marks and evidence of scour and erosional processes are ubiquitous[124] and cannot be taken to be indicative of a shallow water environment. Sediments on the continents, and in particular those of geosynclinal basins (the Lower Palaeozoic geosyncline in Great Britain, the Appalachian geosyncline of North America) contain among them sequences of alternating graded beds and shales with features similar to those from the Iberia abyssal plain, and to those from the Atlantic abyssal plains in general.[73, 154, 216]

Ph. H. Kuenen and others have pointed out[163] that turbidity currents must have been important processes of sedimentation in deposits such as the Alpine Flysch (and by contrast that the Molasse has features typical of neritic and non-marine environments). It is worth while to compare typical Flysch deposits with the sediments from the Iberia abyssal plain.

To the west of the Argentera–Mercantour Massif in the French Maritime Alps lies the Annot sandstone flysch, of Eocene and Oligocene age, which ranges in thickness from 400 to 700 m.[255] The flysch consists of alternating mudstones and sandstones. The mudstones are now compacted and represent pelagic sediments: they contain occasional planktonic Foraminifera. The grain size of the sandstones between is graded, both laterally and vertically and they contain displaced or reworked fragments of rock and displaced Foraminifera. The vertical sequence of a typical rhythmic unit is shown in Fig. 5.7. This should be compared with the descriptions and figures which refer to the Iberia abyssal plain sediments. The features are identical, except for the "current ripple and convolute laminations", and the "sole markings" of the flysch sequence. (The sole markings would be difficult to see in a core only 5 cm in diameter.) One feature of the cores from the abyssal plain either

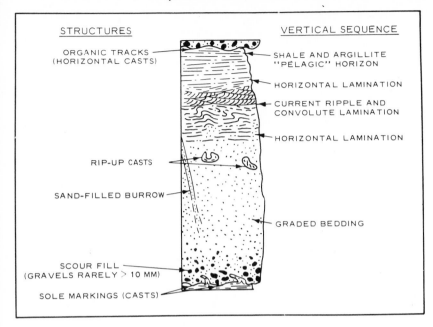

STRUCTURES   VERTICAL SEQUENCE

ORGANIC TRACKS
(HORIZONTAL CASTS)

SHALE AND ARGILLITE
"PELAGIC" HORIZON

HORIZONTAL LAMINATION

CURRENT RIPPLE AND
CONVOLUTE LAMINATION

HORIZONTAL LAMINATION

RIP-UP CASTS

SAND-FILLED BURROW

GRADED BEDDING

SCOUR FILL
(GRAVELS RARELY > 10 MM)

SOLE MARKINGS (CASTS)

Fig. 5.7. A turbidite: sketch of an ideal Annot Sandstone turbidite sequence, showing the variability in vertical stratification and commonly associated sedimentary structures (after D. J. Stanley[255]. Reproduced from the *Journal of Sedimentary Petrology* by permission of the Society of Economic Palaeontologists and Mineralogists).

absent from the flysch sequence, or not described, is the horizontal black bands, ascribed to hydrotroilite, an hydrated iron sulphide.[73] Such sulphide bands are important in understanding the role that bacteria play in the abyssal plain turbidities.[155] Above the graded beds the pelagic sediments often show no sign of a "reducing" environment. By contrast the graded beds of the abyssal plains contain the sulphide bands, and the whole layer which may be black or dark grey when the core is opened, changes colour in the open air of the laboratory, to a light grey and eventually to a brown colour.

D. J. Stanley[88] has described in some detail the lineation and foliation associated with the flysch turbidites; comparable work has not been done in such detail on abyssal plain sediments, but the foliation has been demonstrated by measurements of the anisotropy of magnetic susceptibility. The magnetic susceptibility depends upon the shape and crystallographic structure of a magnetic mineral. If the effect of shape is dominant and there is overall some preferred shape among the magnetic minerals such as magnetite, then the susceptibility will be related to the conditions of deposition;[235] long axes of mineral grains could be oriented in one particular direction, for example. Other physical properties could be used in a similar way — the dielectric constant could be measured, for example.[254]

CHAPTER 6

Movements of the Sea Floor

THE INTEREST

We will ask the question here: "What movements have taken place on the sea floor?" Such a question bears on many problems which are not solved, and among them are those of continental drift and those of the interrelationships of the continents and ocean basins. Two sets of observations will be considered which show (with some certainty) that motion between different parts of the ocean floor is possible both vertically and horizontally. Direct evidence of such motion (which will be reliable) is difficult to obtain; even if surveys of great accuracy were made at intervals between two adjacent land masses and motion of one sort or another were shown to occur it would take some time (thousands to millions of years) to prove that such motions were systematic and not minor fluctuations of little importance. One difficulty which arises in searching for evidence of vertical displacement is choice of reference level. There is good evidence from records of tide gauges, for example, of worldwide general changes in sea-level—eustatic changes, not associated with locally important phenomena leading only to areally restricted changes.[321] An example of recent eustatic change is that recorded by tide-gauges in the past 200 years, and ascribed to melting of polar ice caps. An example of a more local change is the "glacial rebound" documented in Scandinavian countries and Canada, associated with unloading of the crust with retreat of the ice. The one leads to changes in sea-level throughout the world, the other to changes in sea-level only locally. Many factors interact in a complicated way. Sedimentation in basins not only displaces water bath-tub like but the redistribution of mass leads to changes in moments of inertia and the rotation of the earth, and consequently to latitudinal changes in sea-level. Large changes in position of the geographic poles would lead to large changes in sea-level.

There have been many studies of Tertiary, Quaternary and Recent geology which have as their aim the delineation of changes in sea-level, and a summary of such studies has been given by R. W. Fairbridge.[321]

One study which illustrates the techniques and which may be less familiar than many is that of the submarine physiography of the Canadian Arctic and its relationship to crustal movements.[322,332]

One interpretation which can be placed upon the submarine topography of the inlets between the Arctic Islands, northern Baffin Bay and Nares Straits, and the continental shelf north-east of the islands is that it represents a partially submerged Tertiary drainage system. That is, the river-cut valleys of the system were modified by valley glaciation and submerged: subsequently partial emergence has taken place. The margins of Baffin Bay illustrate this (Fig. 3.2), for the submarine valleys of the shelf on both sides of the bay are now U-shaped in profile, have U-shaped hanging tributaries leading into them, and extend into valleys on land occupied by glaciers at the valley head. This interpretation is supported by studies of the sediments of the channels of the Arctic Islands; for example, in some of the cores is found evidence of relatively deep-water fauna overlain by younger, shallow-water fauna. Raised beaches several hundred metres above present sea-level are found on land, and are up to several thousand years old — that is, they represent a post-Pleistocene elevation of the land with respect to the sea.

CORAL REEFS AND GUYOTS

Evidence of vertical movement in the ocean basins comes from studies of coral reef formations and of submerged seamounts of a peculiar nature called *guyots*. No attempt will be made to describe all features of coral reef formations or all theories that have been held which concern the origin of the reefs. We will consider first the superficial elements of coral reef formations.

Coral Reef Formations

Coral reefs are made of accumulations of living and dead assemblages of corals, algae, forams, molluscs and other fauna and their detritus.[56,66,161,294,310] Most post-Palaeozoic fossil and living Recent corals are included with the order Scleractinia,[208] which can be divided into two groups, (1) hermatypic and (2) ahermatypic, differing in the presence or absence of symbiotic dinoflagellates. Rather generally, the hermatypic are the reef-forming corals associated typically with warm, tropical sea water, and with shallow depths, and the ahermatypic are "deep-sea corals", with wider distribution and greater tolerance for a wide range of physical conditions. They cannot be distinguished from one another when dead.[271]

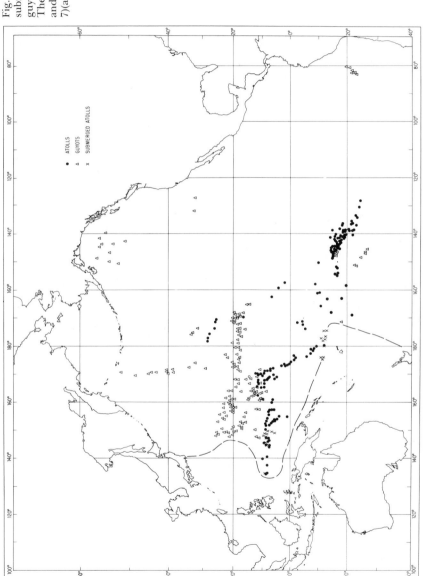

Fig. 6.1. Surface atolls, submerged atolls and guyots in the Pacific Ocean. The dashed line is the andesite line (see Chapter 7)(after H.W. Menard and H.S. Ladd[204]).

ATOLLS

GUYOTS

SUBMERGED ATOLLS

Coral Reef Formations: Hermatypic Corals

The distribution of reefs based on hermatypic corals is controlled mainly by the conditions under which the reef-building corals can live. They thrive best at depths of a few metres below the surface of the water, although they extend down to about 200 m. This arises because of the dependence of algae symbiotic with them upon light for photosynthesis. They seldom grow in water temperatures outside the range 20–30°C or in water of salinity far from that of normal sea water, 35–36 parts per thousand. Because they require oxygen and food, and thrive best in clear water, good water circulation is important; they will withstand exposure to air for a limited amount of time (such as occurs at low tides) but will not withstand prolonged exposure.

The distribution within the seas today of thriving coral-reef formations illustrates their dependence upon these factors; the location of atolls of the Pacific is shown in Fig. 6.1. They are most abundant between 30°N. and 30°S. latitude and are not common on the east sides of the Atlantic and Pacific oceans where the water is cold by comparison with the west sides. The distribution depends too upon the ability of corals to colonize, which will occur if the coral larvae can migrate successfully during the period of time in which they live. This ranges according to species from 2 to 30 days, and success or failure must rest upon the distance from a surface which can be colonized, and upon ocean currents. It is probable that the larvae of invertebrates cannot easily cross the whole open ocean basins, but can migrate from island to island[273] (which they have obviously done). This conclusion extended in application to coral larvae would be supported in two ways: the number of coral genera upon isolated islands such as the Hawaiian or Bermudan islands is relatively small, and the number of genera in the Jurassic coral reefs of Oxfordshire, England, is less than in the regions of maximum development of that time.

This emphasis on distribution may seem unbalanced; such considerations are of great importance in interpretation of geological data to give past climatic conditions. Unique interpretations are rare, and difficulties which can arise are described in the section on cold- and deep-water coral banks.

Reef formations which thrive today can be divided into three classes, for the sake of description primarily but with a relation to genesis also. *Fringing* reefs abut against a rocky coast; *barrier* reefs are separated from the land by a lagoon, and *atolls* are irregular circlets of reef inside which are no islands except perhaps patches of reef. These are illustrated by Figs. 6.2 and 6.3, reproduced from drawings of Darwin.[54] The form

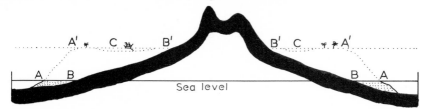

Fig. 6.2. The development of fringing reefs into barrier reefs (after Darwin[54]).
AA—Outer edge of the reef at the level of the sea. BB—Shores of the island.
A'A'—Outer edge of the reef, after its upward growth during a period of
subsidence. CC—The lagoon channel between the reef and the shores of the
now encircled land. B'B'—The shores of the encircled island.

N.B. In this, and the following figure, the subsidence of the land could only be
represented by an apparent rise in the level of the sea.

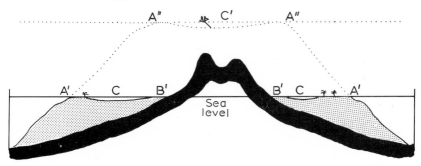

Fig. 6.3. Barrier reefs and their development into atolls (after Darwin[54]).
A'A'—Outer edges of the barrier-reef at the level of the sea. The coconut trees
represent coral-islets formed on the reef. CC—The lagoon channel. B'B'—The
shores of the island, generally formed of low alluvial land and of coral detritus
from the lagoon channel. A"A"—The outer edges of the reef now forming an
atoll. C'—The lagoon of the newly formed atoll. According to the scale, the
depth of the lagoon and of the lagoon channel is exaggerated.

and structure of an atoll will be illustrated by considering Bikini and
Eniwetok atolls.[2,59,66,231−2]

Bikini and Eniwetok Atols

Bikini and neighbouring atolls and guyots lie in the western Pacific
Ocean, a part of the Marshall Islands (Figs. 6.4 and 6.5). Bikini rises
from the deep ocean floor at about 6000 m depth. The lagoons vary in
shape from circular to rectangular and are surrounded by a shallow
submerged terrace and then by coral knolls. The seaward edge of the
reef is bordered by another submerged terrace, rather shallower
than the lagoon terrace and the edge of this terrace may fall rather

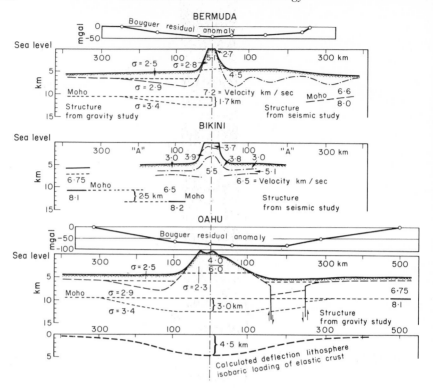

Fig. 6.4. The structure of Bermuda, Bikini and Oahu deduced from gravity and seismic observations. σ is the observed density (g/cm³), and other figures represent thickness in kilometres or compressional wave velocities in km/s (after M. N. Hill[136] drawn from observations by R. W. Raitt, G. P. Wollard and others).

steeply for 40 m or more. The outer slope is quite steep to about 400 m and has an average value of about 45°; below 400 m the slope eases until it merges with the floor of the deep sea. The debris which surrounds the reef near the top is made principally of coral debris, algal fragments and forams, and the amount of coarse fragments decreases down the slope until Foraminifera predominate. Finally, "red clay" replaces the Foraminifera, below the solution depth of calcium carbonate.

Drilling on Eniwetok and Bikini showed that the upper few hundred metres of the atolls are detrital, organic limestones which accumulated in an environment similar to that of the uppermost parts of the atolls at the present time. The drilling at Eniwetok penetrated an igneous rock, olivine basalt, at a depth of 1·27 km, and dredgings on the sides of Bikini showed that there, too, basalt underlies the limestone of the drill-holes.

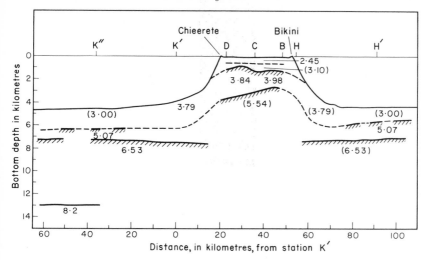

Fig. 6.5. Cross-section through Bikini Atoll, showing the structure deduced from seismic observations. The figures show compressional wave velocities in km/s, and where bracketed are uncertain (after M. N. Hill[136] drawn from observations of R. W. Raitt).

Results of Raitt's seismic study at Bikini are shown in Fig. 6.5, and comparison with the drill-holes suggests that his layer with velocity 3·98 km/s is basalt and the upper two layers are calcareous deposits. Detailed studies by S.O. Schlanger of the subsurface geology of Eniwetok atoll based on the cores from the drill-holes are most interesting. He has shown that the geography of the atoll changed with time – the earliest reef forming on the volcanic base on what is now the southeast side. The tops of both Bikini and Eniwetok were raised 200–300 m above sea-level several times in the time interval Miocene to Recent. This is shown by the nature of the rocks in the cores. Unaltered sediments are unconsolidated skeletal remains made of aragonite. Upon emergence near-surface water dissolves the aragonite and reprecipitates calcium carbonate as calcite, causing lithification of the formerly unconsolidated material. Subsequent subsidence is indicated by the presence of more unconsolidated aragonitic sediment on top of the lithified material. The times of successive emergence and submergence at both atolls were the same.

The igneous rocks from the dredgings off Bikini are in part composed of fragments which result from explosive activity and in part vesicular basalts. They are altered; for example, olivine phenocrysts are recognized mainly by shape, because the olivine has altered to "chlorite and serpentinous material", and sometimes to "bright red iddingsite". Macdonald's description (in reference (66), pp. 120–4) is similar to

Matthew's description of igneous rocks from the Iberian abyssal plain (Chapter 7).

The evidence seems strong that the two atolls were once subaerial volcanic islands which began to subside in the Miocene – the age of the limestones above the basalt beneath Eniwetok; the total subsidence has been approximately 1 km. Coral reefs have maintained themselves at sea-level, keeping pace with the subsidence and forming the capping to volcanoes.

Guyots

The guyots of the Marshall Islands and of the Mid-Pacific Mountains[107] are flat-topped seamounts which rise from the floor of the deep-sea

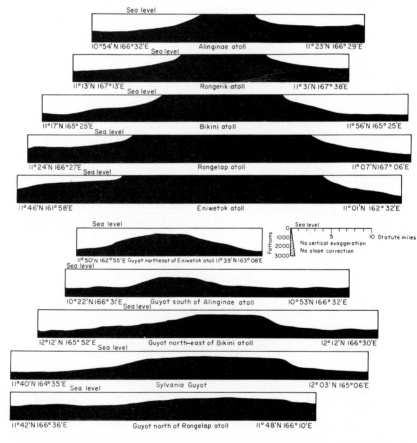

Fig. 6.6. Cross-section of atolls and guyots. The peripheries of the guyots are rounded, those of the atolls comparatively sharp (after K. O. Emery, J. I. Tracy and H. S. Ladd [66]).

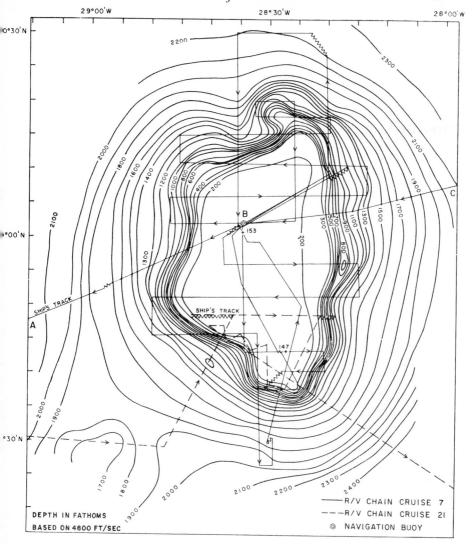

Fig. 6.7. Great Meteor Seamount. Bathymetric contours at 100-fm intervals, with sounding lines (after R. M. Pratt[227]).

to depths in the range 1000–2000 m from the surface of the sea. They have rather flat tops, but differ from atolls in that the highest point of a guyot is near the centre of the top, the highest point of atolls being at the edges of the atoll; the edges of guyots are rounded, those of atolls sharp (Figs. 6.6 and 6.7). The rocks on the tops of some guyots are a mixture of rounded basalt pebbles and sandstones, together with a fauna including reef-corals, rudistids, and Mollusca; the fauna is Cretaceous in age. The reef corals suggest that the water in which they formed was shallow, and the rounding of a number of the basalt pebbles suggests that erosion by waves occurred at some time, which led to the shelf-like top.

SUBSIDENCE AND ELEVATION

The evidence is this. Reef corals are restricted in habitat to a shallow depth and to tropical or semi-tropical seas. They live now on coral atolls, fringing reefs and barrier reefs, and drills show their presence to depths of more than 1000 m below present sea-level under the atolls. Basalt is found beneath the reef deposits of Eniwetok atoll and outcrops on the lower slopes of the atolls. Basalt outcrops too on the tops of guyots, but a Cretaceous fauna is also found there which can have lived only in shallow water. Evidence of erosion of the tops of guyots by the action of waves is strong because of the shape of the tops and the rounding of the pebbles.

Hamilton suggested that the guyots are submerged volcanoes, which were once at sea-level but subsided. This is supported by the evidence of the atolls, which are themselves built up upon volcanoes. However, although it is tempting to try to ascribe the formation of guyots to subsidence at a rate downwards faster than the coral reefs could develop upwards, there is little evidence that the guyots were ever atolls,[199] notwithstanding the reef–coral fauna. In particular, their shapes are more like submarine volcanoes than atolls, with no steepening of slope near the upper parts of the structure. Of course, the steep uppermost slopes of modern atolls may be due to fast upwards growth in post-Pleistocene time.[107] A mechanism is needed which explains how volcanoes built above sea-level and eroded by wave action sank. The guyots are found concentrated in particular regions, the most important being the west central Pacific Ocean, so that world-wide overall changes in sea-level are not necessarily required.

Among the phenomena which could lead to the present level of the tops of guyots are: sinking of individual guyots (then volcanoes), sinking of groups of volcanoes, sinking of a whole region and overall rise in sea-level with respect to the ocean floor.[199] The sinking of individuals or of groups is unlikely because the evidence of subsidence in the form of a moat in the sea floor found around the Hawaiian Islands, for example (Chapter 7), is not in any way universally found with the individual guyots or with the groups of them. A general rise in sea-level equal to one-quarter of the present ocean depths might have led to extensive temporary flooding of the continents before isostatic equilibrium was

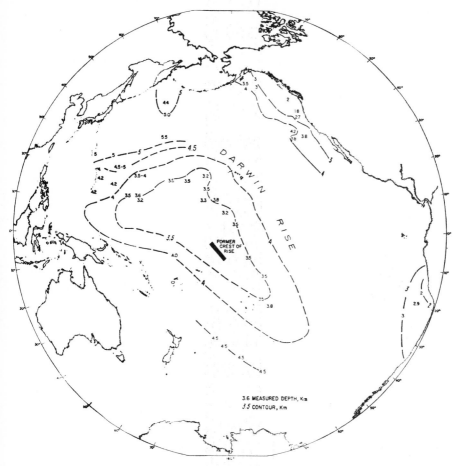

Fig. 6.8. Paleobathymetry of the Pacific Basin based largely on the local relief of guyot platforms. Paleodepths off South America and the Gulf of Alaska probably are not contemporaneous with the Darwin Rise. Position of former crest of rise based on faulted topography and relief of guyots (after Menard[199]).

re-established; this would have occurred in Cretaceous and later times and evidence for it is poor. The rate of out-pouring of water from the mantle would also have been much greater than is likely.

Menard[199] has attempted to reconstruct the bathymetry of the southwest part of the Pacific Ocean as it was at one time by contouring the elevation above the present sea-floor of the "shelf-break" of the tops of guyots and the basal platforms of the atolls upon which reef-corals developed (where this is known or can be estimated) (Fig. 6.8). The contours suggest that a mid-ocean ridge (called the Darwin Rise) comparable to the present mid-ocean ridges existed, and has since subsided. This leads naturally to the suggestion that the formation of the guyots was on a regional scale, whole areas being uplifted and the tops of elevated volcanic peaks being eroded by wave action; subsequently, the volcanoes sank with the demise of the Darwin Rise. The development of guyots rather than atolls will presumably be related not only to the speed of subsidence and the success or failure of the organic growth to grow upwards sufficiently quickly, but also to the environmental conditions which would allow or prevent the development of the (hermatypic) reef corals.

The abundance of guyots in the western Pacific Ocean and their rarity in other oceans could reflect the different stages of development of the ocean floor throughout the world. The Darwin Rise has risen and sunk, but the East Pacific Rise and the Mid-Atlantic Ridge have only risen, so far as is known, except no doubt in a few restricted areas. Great-Meteor Seamount in the Atlantic Ocean (Figs. 6.7 and 6.9) has a guyot-like shape; its top is now only 300 m below sea-level, and it would not be absurd to suppose that this guyot owes its shape to erosion at a time of lower sea level, rather than to local or regional sinking of the sea floor.

We may ask, if subsidence of reef-formations has occurred, has elevation not happened also? It has, and occurrences of elevated reef formations may be found in the Fiji Islands, for example. The authors of the Bikini monograph[66] point out that in Melanesia, Tertiary reef limestones outcrop extensively, but outcrop little or not at all to the east (in Micronesia); this, they suggest, is of regional significance dividing this part of the Pacific into two, one in which subsidence has been the rule, the other where elevation has been the rule.

The vesicularity of basalts from dredge hauls has been taken as evidence of outpouring in relatively shallow water. This need not be so; vesicular basalts are found under 4 km of water in the Atlantic Ocean (Chapter 7).

Cold- and Deep-water Coral Banks: Ahermatypic Corals

The main purpose of this chapter is to illustrate the probability that there have been large movements of the earth's crust. However, it is convenient here to point out that the ahermatypic corals are capable of building ramified colonial skeletons and do so in relatively deep, cold water off western Europe. Teichert has described such banks off Norway, and points out that if found to be fossil there might be little to distinguish them from the coral reefs of the hermatypic corals. Quite the wrong conclusions could be drawn about the environment.[271] In

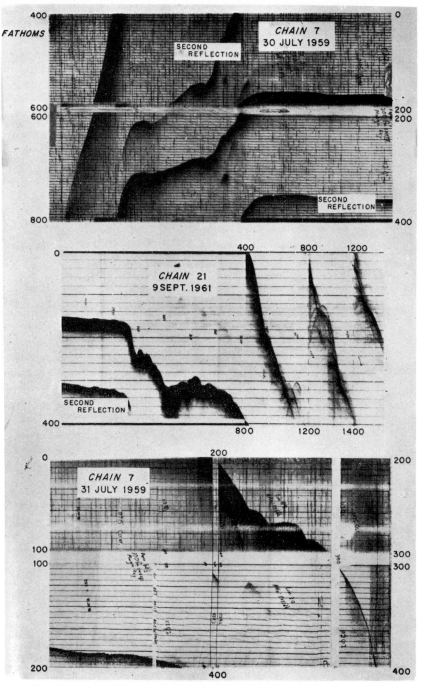

FIG. 6.9. Great Meteor Seamount. East–west cross-section with two different vertical exaggerations (after R. M. Pratt[227]).

the particular case of the Norwegian banks the environment would perhaps be supposed to be: water depth 30–50 m, temperature greater than
20°C, location-open shelf. In fact the corals are found at depths in the
range 200–300 m, in water with temperature near 5°C associated with
the bays and fjords of a complex coast. Other fauna and flora or sedimentary structures may be diagnostic, of course, of temperature or
depth – the presence or absence of oolites and algae for example. The
difficulties of deducing the environment of Palaeozoic corals is made
even harder because these corals are now extinct.

LATERAL DISPLACEMENTS: MAGNETIC ANOMALIES IN THE EASTERN PART OF THE PACIFIC OCEAN

The Physiography and Magnetic Anomalies

There is thus good evidence of vertical displacements of at least
2 km. Investigations in the eastern part of the Pacific Ocean show that
lateral displacements which are in total 1450 km have occurred, and the
studies which showed this, showed too a peculiar grain in the oceanic
crust.

The eastern part of the Pacific Ocean floor between 130°W. longitude
and the North American continent has a varied character. The Baja
California seamount province lies to the south of the area and to the
north is the deep plain of the north-east Pacific; further north still is
the Ridge and Trough province. These physiographic provinces are
distinct and different one from the other and there would be no reason
to suppose, if topography were used as a guide, that the crustal structure
associated with each province is not as distinctive as the topography.
A number of major faults associated with escarpments cross the area,
among them being the Murray, the Mendocino and the Pioneer. The
continental shelf and slope lie to the east; the San Andreas fault runs
through California nearly parallel to the coast and this fault is still
active. Its sense is right handed, that is, an observer on one side of the
fault sees that displacements on the other side are to his right.

Magnetic surveys made at sea in this area show two major features,
a north–south lineation of anomalies, broken at the Murray and the
other escarpments.[190,282] A comparison of east–west profiles of magnetic
anomalies found to north and south of the escarpments was made by
Vacquier, Raff and Warren. Profiles between any two fault zones are
similar one to another but profiles across any fault zone are similar only
if one is displaced laterally. The displacements must be left-lateral

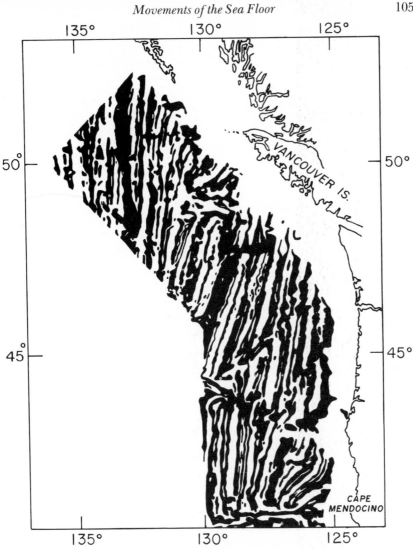

Fig. 6.10. Map of magnetic anomalies north of the Mendocino escarpment
(after A. D. Raff and R. G. Mason[230]).

across two faults, 265 km in the case of the Pioneer and 1185 km in the
case of the Mendocino, whereas that associated with the Murray fault is
right-lateral and 150 km in magnitude (Fig. 6.11). The area north of
the Mendocino escarpment is shown in Fig. 6.10. Mason has pointed
out that the area from 41°N. latitude to 49°N. latitude looks like a jig-saw
puzzle of angular pieces of crust that have slipped and rotated with

respect to one another. The north–south lineations disappear at the continental margin, and no certain extensions of the oceanic faults on the continent to the east have been found. This is illustrated by Fig. 6.11. Faults near shore with the same trends as the Mendocino and Murray faults, have smaller displacements in the opposite sense. The San Andreas fault is still active and there may have been movement of 560 km along it since the Jurassic[135] whereas the faults such as the Mendocino

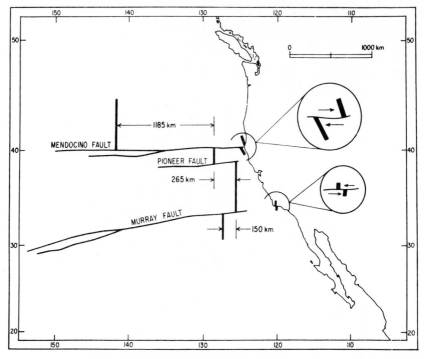

Fig. 6.11. Fracture zones in the east Pacific. The heavy vertical line represents the axis of a north–south magnetic anomaly which was continuous before slipping started (after H. W. Menard[198]).

are seismically inactive west of 130°W. longitude.[282] The presence of a complicated fault system will make more difficult the recognition of the extension of the oceanic features on to the continent.

M. D. Fuller[90] has attempted to trace the east–west fracture zones of the east Pacific across the North American continent, by cross-correlation of east–west profiles of the magnetic field from north to south. He has concluded that the fracture zones do continue into the continent, and suggests that it will be worthwhile to look for similarity between fracture zones of the Pacific Ocean and the Atlantic Ocean.

The Cause of the Anomalies

The most obvious cause of the magnetic anomalies is a linear array of magnetic bodies of rock situated in the oceanic crust; plate-like bodies in the upper part of the oceanic crust, which could perhaps be lava flows, or dyke-like bodies through the whole oceanic crust would yield magnetic anomalies similar to those which are observed.[230] This may be the whole truth; if it is, the problem changes, for the composition and the cause of the long and linear magnetic bodies of rock must be sought. An alternative explanation has been proposed[284] in connection with anomalies over mid-ocean ridges, and this will be discussed in Chapter 8.

CHAPTER 7

Igneous Rocks
of the Ocean Basins

Igneous rocks make up the major part of the crust of the ocean basins. The melts from which they crystallize are generated at least partly within the mantle. The oceanic crust is thin by comparison with the continental crust and the igneous rocks on the ocean floor will consequently be a less contaminated product of mantle rocks. Hence they are important in any study of the composition of the earth. The forms they assume – as volcanoes or intrusions, and the history of the structures tells us about the development of the oceanic crust. They break down on weathering into a variety of minerals which form part of the sediment of the ocean floor. To study igneous rocks is then important.

Evidence of their nature comes from oceanic islands and from samples dredged from the ocean floor. Full understanding requires also experimental studies in the laboratory, and comparison with igneous rocks of the continents. The ocean basin features of which the geology is known best are the islands, but these occupy a small fraction of the total area of the sea floor. Unfortunately, the number of dredge hauls in the remaining part of the ocean basins is small so that the data about igneous rocks from which conclusions may be drawn is meagre. Many of the igneous rocks of the ocean basins are basic or ultrabasic and a short account of some topics concerning such rocks in general is presented in the next section.

BASIC AND ULTRABASIC IGNEOUS ROCKS

Acid rocks such as granites, which are found principally on the continents, can be distinguished by their chemistry and mineralogy from basic rocks, such as gabbros and basalts. The content of silica SiO_2, in a chemical analysis will be near 70% for a granite or its fine-grained equivalent rhyolite, and the principal minerals will include quartz, orthoclase feldspar, sodium-rich plagioclase feldspar and mica. The content of silica in a basic rock such as basalt will be near 50%, and the principal minerals will include calcium-rich plagioclase feldspar and magnesium and iron bearing minerals, often pyroxenes and olivines. Gabbros occur as coarse-grained plutonic rocks, and basalts as fine-grained volcanic lavas or thin intrusions. Variants within this group are defined in terms of the type of pyroxene or the presence or absence of either pyroxene or olivine. Ultrabasic rocks or ultramafic rocks are poorer in silica than basic rocks, containing about 40% by weight, and are composed mainly of olivine, with lesser amounts of pyroxenes and plagioclase feldspars. Dunite is wholly olivine. The rock eclogite is of rare occurrence in the ocean basins but may be of great significance. It consists of a pyroxene (omphacite) and a garnet; the omphacite is rich in Na_2O and Al_2O_3 by comparison with the pyroxenes of ordinary igneous rocks. The garnets are also characteristic, being calcium rich varieties of the three typical garnets almandine, grossular and pyrope. Garnets have the general formula:

$$R_3^{2+} R_2^{3+} Si_3O_{12}$$

In this case R^{2+} may be Ca^{2+}, Mg^{2+}, Fe^{2+}, and R^{3+} may be Al^{3+}.

Eclogites are chemically equivalent to gabbros, and this may be seen in equations of the following type, in a formal sense.

Gabbro

$$(CaAl_2Si_2O_8 + NaAlSi_3O_8) + Mg_2SiO_4 + CaMgSi_2O_6 =$$

Labradorite:	Ca-rich	Olivine	Diopside:
	Plagioclase		Pyroxene

Eclogite

$$CaMg_2Al_2Si_3O_{12} + (NaAlSi_2O_6 + CaMgSi_2O_6) + SiO_2$$

Garnet	Omphacite	Quartz

This equivalence is important for us not only in this special case but because it illustrates how a particular chemical analysis derived from a particular mineral assemblage could also arise from a rock with completely different mineralogy. The chemical composition of a rock will be reflected in its mineralogy, and a rock which is rich in quartz will probably show a high percentage of SiO_2 on chemical analysis. However, the discussion of eclogites shows that the chemical composition does not uniquely determine the mineralogy. We can, given the results of chemical analysis, rearrange the figures to yield a set of minerals which might be present, and the assemblage of minerals so calculated is the normative assemblage; the assemblage actually present in the rock is the modal assemblage. The significance will be seen in the discussion of basalts. Rules exist for calculating the norm, called the CIPW rules (after the authors), and everyone uses these rules so that normative calculations are compatible one with another.[12] The normative minerals (rather than the modal minerals) are significant if, for example, a rock has suffered metamorphism since its original formation. The modal minerals now may be certainly different from those originally formed, but the norm calculated from the chemical analysis may approach the original mineralogical composition. In this way H. D. B. Wilson and others[296] were able to recover the original characteristics of the metamorphosed Keewatin lavas of the Canadian Shield, and consequently compare them with younger, less altered volcanic rocks of continents and ocean basins.

A whole suite of igneous rocks may crystallize from any one melt. Consequently the question might be asked whether a melt of any particular composition could generate basalts, andesites and rhyolites in turn—the suite becoming richer in SiO_2 successively. Granites are known in the ocean basins on the Seychelles on the one hand in great quantity, on Ascension Island on the other hand in minor quantity. If in both cases it is reasonable to envisage a reservoir of molten basaltic liquid beneath the ocean floor which can be supposed to generate the granite, then it is unreasonable (from this viewpoint) to suggest that the Seychelles are part of a micro-continent. A second example can be taken from Hawaii, where the lavas poured out change in composition with time. Fractional crystallization in elegant simplicity was studied by N. L. Bowen.[26,27] Suppose, for example, we take an atomic substitution series of minerals such as the plagioclase feldspars, which range from a composition of $NaAlSi_3O_8$ to $CaAl_2Si_2O_8$, CaAl substituting for NaSi in successively greater proportion. The first mineral to crystallize from a melt of a composition which is a mixture of the pure end members is calcium–rich by comparison with the melt itself. If these first-formed crystals are removed from the system the overall composition and consequently the final solids precipitated are sodium-rich by comparison with the original melt. In this way the minerals derived from a melt of one composition have been separated into a sodium-rich and a calcium-rich fraction. Similarly, in the olivines ($Mg_2SiO_4 - Fe_2SiO_4$) magnesium-rich minerals are precipitated first. If removed from the system the final minerals to precipitate are iron-rich. The lavas of Hawaii exhibit the change in the plagioclase feldspars. The lavas poured out successively richer in SiO_2 have plagioclase feldspars successively richer in sodium. The experimental studies of Bowen (and many others) suggests that this arises by some process of fractional crystallization or fractional melting, not because parts of the mantle of different compositions are being tapped at different times. Iron enrichment in olivines, together with sodium enrichment of plagioclase feldspars and silica enrichment of the rocks overall is seen in many basic intrusions where the cooling liquid was trapped in some sort of container or magma chamber. A striking example is the Skaergaard intrusion of east Greenland,[286] which forms a layered series of rocks. Such layered intrusions must be common beneath the ocean floor and will have acted as sources for some of the lavas which are poured out. Muir, Tilley and Scoon think that the particular characteristics of olivine crystals found in basalts of the Mid-Atlantic Ridge indicate the presence of such an intrusion.[209]

Basalts

Basalts as rock types consist principally of plagioclase feldspar and pyroxene, and may carry a variety of other minerals, olivine, nepheline ($NaAlSiO_4$), quartz and iron oxides among them. The plagioclase feldspar is (usually) relatively calcium-rich. Other considerations apart, their chemical analyses would allow derivation from eclogite directly, or by partial melting of an ultrabasic rock such as peridotite. They fall into two groups, the *alkali basalts* and the *tholeiitic basalts* or tholeiites. These two groups are chemically and mineralogically dissimilar and questions asked about them are of the following type. (1) Is one group derived from the other, and if so, which? (2) If not, do they have a common parent rock (or melt)? (3) Where within the crust or mantle is the magma generated which leads to the basalts observed at the surface?

The alkali basalts are olivine-bearing and quartz-free. They are associated with a series of rock types which may have originated by differentiation from the alkali basalts, and this series shows a progressive enrichment in SiO_2, Na_2O and K_2O. Consequently the minerals are sodium-rich, and sodium-rich plagioclase feldspars are found associated with pyroxenes because the SiO_2 content is too low (relatively) for minerals more SiO_2 demanding, or quartz itself, to crystallize. This leads to rocks such as the oligoclase–basalts, or mugearites, and the nepheline-bearing intermediate rock, phonolite. Normally oligoclase (sodium-rich plagioclase) would not be associated with pyroxenes, and nepheline is the product of a melt rich in Na_2O and relatively deficient in SiO_2. The tholeiitic basalts are basalts in which quartz may be found in small amounts and which are, on the whole, olivine-free (but see below). They are associated with rocks which show a progressive enrichment in SiO_2, but not such a conspicuous enrichment in Na_2O and K_2O as in the series associated with the alkali basalts. The tholeiitic series includes the dacites and rhyolites (in which, typically, Na_2O is accommodated as a plagioclase feldspar, not as nepheline).

However, these descriptions of tholeiitic and alkali basalts have been only in terms of the minerals actually present in the rock—the modal minerals. Even if lava is saturated or oversaturated with respect to silica (SiO_2), olivine may appear in the rock itself because olivine crystallizes early and may not be reabsorbed before crystallization is complete[275] (that is to say olivine may not appear in the norm but it will appear in the mode; had there been time, perhaps the olivine would not have appeared in the mode, i.e. it would not appear visibly in the rock itself). This means that our quartz-bearing tholeiitic basalt may properly be described by an observer as an olivine basalt in the sense that it is olivine-bearing. Consequently, it may be more convenient to consider the normative components, because these are easily recalculated from chemical analyses.

Yoder and Tilley divided tholeiitic and alkali basalts into five groups, according to the presence or absence of one or more of normative quartz, hypersthene (a pyroxene), olivine and nepheline.[309] The two extreme groups, which we can take as indicative of the remainder are tholeiite (oversaturated) with normative quartz and hypersthene, and alkali basalt, with normative olivine and nepheline. Both tholeiitic basalts and alkali basalts are present among the rocks of the ocean basins on Hawaii for example: two chemical analyses are presented in Table 7.1 to allow comparison between them. These have been taken from the analyses of Yoder and Tilley (reference (29), Table 2, Numbers 7, 20 respectively). The tholeiite was described as a typical tholeiite; olivine phenocrysts occur in the rock but it is essentially saturated having only 0·30% quartz in the norm. The (modal) plagioclase has composition An_{52-49}. The alkali basalt was described as a typical alkali basalt, "rich in granular olivine as microphenocrysts". Table 7.1 shows essential differences between the two types. The tholeiite has higher SiO_2, lower ($Na_2O + K_2O$), higher normative quartz, lower (none) normative olivine.

The chemical similarity of eclogites as a whole to basalts as a whole has been mentioned already. Yoder and Tilley point out that for every type of basalt a corresponding eclogite can be found, and also that an alkali basalt could be generated from an eclogite by subtraction of the garnet phase, and a tholeiite by substraction of the pyroxene (sodium-rich omphacite). If the chemical analyses of the garnet phase is recalculated in terms of normative minerals, anorthite, hypersthene and olivine result; the normative minerals of the omphacite are principally albite and diopside, with lesser amounts of nepheline and anorthite. This suggests that one mechanism for the generation of

Table 7.1. *Chemical analyses and norms of a tholeiite and an alkali basalt*

1. Chemical analyses   2. Norms

	(1) Tholeiite	(2) Alkali basalt		(1) Tholeiite	(2) Alkali basalt
SiO_2	51·18	46·53	Quartz	0·30	—
Al_2O_3	14·07	14·31	Orthoclase	2·22	5·28
Fe_2O_3	1·35	3·16	Albite	20·96	20·04
FeO	9·78	9·81	Nepheline	—	2·20
MnO	0·17	0·18	Anorthite	26·13	23·63
MgO	7·78	9·54	Diopside	22·04	20·89
CaO	10·83	10·32	Hypersthene	22·44	—
Na_2O	2·39	2·85	Olivine	—	18·48
K_2O	0·44	0·84	Magnetite	1·86	4·53
H_2O^+	0·10	0·08	Ilmenite	3·95	4·41
H_2O^-	0·15	nil	Apatite	0·34	0·67
P_2O_5	2·10	0·28	Calcite	—	—
TiO_2	0·05	2·28	Rest	0·11	0·14
Cr_2O_3	n.d.	0·06			
SrO	n.d.	n.d.	Total	100·35	100·27
NiO	n.d.	n.d.			
CO_2	n.d.	n.d.			
SO_3	n.d.	n.d.			
Total	100·40	100·24			

3. Quantities of interest

	(1) Tholeiite	(2) Alkali basalt
$Na_2O + K_2O$	2·83	3·69
$\dfrac{FeO + Fe_2O_3}{MgO + FeO + Fe_2O_3}$	0·589	0·576
Total feldspar	49·31	48·95
Per cent An	53·0	48·3

Notes on Table
1. The tholeiite (1) is No, 7, Table 2 of Yoder and Tilley.[13] (Muir, Tilley and Scoon,[9] 1957, p. 244, Cambridge No. 57358, Analyst: J. H. Scoon.
2. The alkali basalt (2) is No. 20, Table 2 of Yoder and Tilley.[13] Cambridge No. 65992. Prehistoric flow North of Keauhou. No. FM10 collected by G. D. Fraser Analyst: J. H. Scoon.)
3. n.d. not determined
4. Norms calculated after CIPW conventions.

basalts within the ocean basins may be by phase change from eclogite, and, according to pressure and temperature, either a tholeiite or alkali basalt will result. However, the stability fields of basalt and eclogite imply

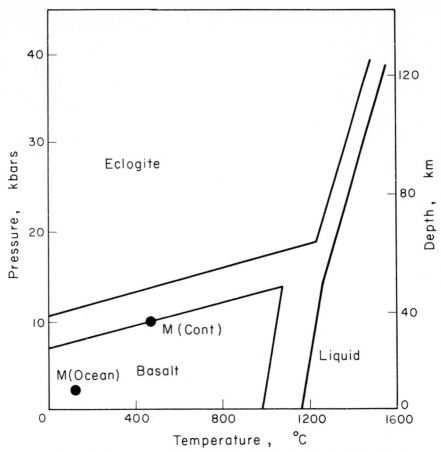

Fig. 7.1. The pressure–temperature relationships for basalt and eclogite. Depth scale obtained with density 2·73 g/cm³, 0 to 37·5 km, and 3·33, from 37·5 km. This is the interpretation by Yoder and Tilley[309] of their experimental results. The lines enclose the transformation zone. The solid circles mark the P–T condition of the Mohorovičić discontinuity beneath continents and oceans, deduced by Lovering[180] (after H. S. Yoder and C. E. Tilley[309]).

that eclogite cannot be stable in the uppermost mantle beneath ocean basins; it may also be too radioactive (and therefore too heat productive) for the suggestion to be reasonable (see Chapter 8).

Basalts are ubiquitous within the ocean basins, and seismic studies suggest that basic rocks may form the major part of the oceanic crust. However, it is possible that the oceanic crust is made of altered upper mantle rock. Partial melting of ultrabasic rocks such as peridotite could produce basalts, as a chemical change rather than a phase change. To postulate such an ultrabasic rock as a constituent of the upper mantle is attractive.

They occur in many fold mountain belts (the so-called Alpine type) such as the Appalachian Mountains and the South Island of New Zealand;[46] they occur too on St. Paul's Rocks in the Atlantic Ocean, and these are considered later. The olivines are often altered to serpentine, and one reaction that might give rise to such a change is

$$5Mg_2SiO_4 + 4H_2O = 2(OH)_4Mg_3Si_2O_5 + 4MgO + SiO_2.$$

Olivine　　　Water　　　Serpentine　　　　Brucite　　Silica

Brucite is not associated in large quantity with the serpentinized olivines, so that this reaction must be accompanied by a demand for its removal, say in aqueous solution. Another reaction, not leading to the formation of brucite but needing SiO_2 is[15]

$$3Mg_2SiO_4 + SiO_2 + 4H_2O = 2(OH)_4Mg_3Si_2O_5.$$

Experimental studies of the system $MgO-SiO_2-H_2O$ at pressures up to 3 kilobars (equivalent to a depth in the earth of about 10 km) and temperatures up to 1800°C have been made.[28] These show that no melt which would subsequently crystallize as serpentine can exist below 1000°C. At 1000°C a magnesian–olivine melt containing water vapour will yield olivine crystals with water vapour between them. Serpentine and brucite (MgO) form at 400°C, provided the water is still available; that is, serpentine forms in the manner suggested by the equation above. Olivine cannot be converted to serpentine above 500°C in any way. These experiments suggest that the lower limit within the earth above which serpentinization is possible, below which it is impossible, is controlled by the 500°C isotherm. If serpentinized ultrabasic rock does form the main oceanic crust then the Mohorovičić discontinuity is the location of this isotherm. One factor which is important in reactions such as those just given is the change in volume which occurs. If a large mass of olivine-rich rock is serpentinized underneath the ocean basins the volume changes which result will affect the (observed) topography and structure of the oceanic crust. Furthermore, the elastic properties are altered so that the velocity of propagation of seismic waves is lower than in the unaltered rock.[21]

IGNEOUS ROCKS OF THE OCEAN BASINS

Igneous rocks of the ocean basins have been studied on the oceanic islands, and as samples dredged from the sea floor. A brief account of the geology of a number of islands follows, the islands chosen to be representative of special situations. Thus Iceland sits astride the Mid-Atlantic Ridge. The Seychelles form an anomalous mass which resembles a continental fragment. St. Paul's Rocks are composed of dunites, ultra-basic rocks which could form a part of the upper mantle. A number of questions arise in the study of the geology of oceanic islands: what are the igneous rocks present, what are their origins, and what is their relation-ship to other igneous rocks of the ocean basins and continents. An account of oceanic islands has been presented recently by J. Tuzo Wilson.[298]

OCEANIC ISLANDS

Iceland

The particular interest of Iceland is that it lies on the Mid-Atlantic Ridge and that, considering the Tertiary igneous rocks of the island, it is a part of the volcanic province of the Tertiary which included the British Isles, east Greenland and Iceland; this is the Thulean or Brito-Arctic province. Also, it is the most productive volcanic centre which can be directly observed in the world (it is difficult to judge how much lava is poured out on the sea floor every year). Lavas both of tholeiitic and alkali basalt affinity are present in eastern Iceland[25,288-9] and are associated with minor quantities of rhyolites and andesites; this is typical of the whole of the Thulean Province.[279] A graben (a part of the crust lower—stratigraphically or physically—than the surrounding crust and bounded on its sides by faults) is supposed to run through the central part of Iceland from south-west to north-east,[111] and it has been sug-gested that this is a continuation of a rift valley associated with the crest of the Mid-Atlantic Ridge. The oldest rocks exposed are of Tertiary age. Some of the lavas are reversely magnetized;[143] that is, the direction in which they are now magnetized is approximately opposite in direction to that of the earth's magnetic field at present, which is rather close to that of the field in the Tertiary.[51] Moreover, adjacent lava flows may be magnetized in opposite directions. Heat-flow through the crust in Ice-land is in the range 2–3 μcal/cm^2 sec,[272] higher than the average values on the ocean floor or on continents, but about the same as is found in numbers of places on mid-ocean ridges.

The structure of the crust beneath Iceland is given below.[13]

	Thickness km	Compressional wave velocity km/s
(Flood basalts)	2·1	3·69
	15·7	6·71
	10·0	7·38
(Mantle)		8·2

Total crustal thickness: 27·8 km

These results are interesting because they show a low velocity associated with the uppermost basalts, and a rather high velocity lower in the crust of 7·38 km/sec. The low velocity is comparable with that observed in the upper parts of the oceanic crust, and is some evidence that they too may be composed of volcanic rock such as basalt. The high velocity lower in the crust could arise from a mixing of ultrabasic mantle rocks with basic crustal rocks. However, such velocities are certainly found elsewhere; although the crustal structure beneath Iceland is more like that measured elsewhere on the Mid-Atlantic Ridge[77] some continental crusts show a very similar distribution of velocities in their lower parts.[76]

Bodvarsson and Walker[25] have shown that Iceland has extended laterally in the past few tens of millions of years at a rate comparable to that required for the continents of Europe and Africa and the Americas to have drifted apart in the past few hundreds of millions of years. The evidence is geological; older rocks are to be found on the eastern and western margins of Iceland, younger rocks in a central belt which is approximately colinear with the Mid-Atlantic Ridge crest. The sides have moved apart and a whole multitude of little dykes pushed in. Since the early Tertiary, the east and west parts of Iceland have separated by 400 km. Now multiple dyke injection provides a neat mechanism for continental drift — a dyke just shoulders Europe and America aside a little more, periodically. Unfortunately, the crust is not strong enough to withstand such pushing without crumpling, for which there is little evidence, and so it appears that dykes *follow* establishment of tension in the crust by motion in the mantle. This is considered again in Chapter 9.

St. Paul's Rocks

St. Paul's Rocks are unlike typical volcanic oceanic islands. They are a small group of islets which lie on the Mid-Atlantic Ridge near to the equator. The rocks are dunites, deformed and ground-up, and fall into two groups, the one an assemblage of olivine and amphibole, the other an assemblage of olivine and a pyroxene enstatite.[274] There are no volcanic rocks. Ultrabasic rocks and serpentinites are often found on continents in areas where there has been deformation,[279] and St. Paul's Rocks are in a region of intense fracturing of the Mid-Atlantic Ridge. The serpentinization and the presence of amphiboles shows that at some stage in their history the rocks had access to water; this is considered further in Chapter 8.

Ascension Island

Ascension Island stands in contrast to St. Paul's Rocks in being more nearly a "typical" volcanic oceanic island. It lies on the Mid-Atlantic

Ridge in the South Atlantic Ocean and is a volcanic cone which rises to 858 m above sea-level from a ridge at a depth of about 3000 m. Ascension Island is a member of the alkali basalt volcanic association and petrologically is similar to Samoa in the Pacific and St. Helena in the Atlantic.[279] The rocks are predominantly olivine bearing basalts associated with alkaline trachytes and similar lavas. Rhyolites and obsidian, a glass of granite composition, are also found, and so too are xenoliths of coarse-grained igneous rocks of a variety of compositions, including among them both granites and gabbros.

The association of such a variety of rock types is difficult to interpret. The presence of granite xenoliths led to the conclusion that a "continental" complex of rocks underlies the Mid-Atlantic Ridge,[53] but this is not now thought at all probable, and they must be regarded as the result of differentiation from a basic melt. This would lead too of course to the rhyolites and obsidians—the extrusive volcanic representatives of the granites. The difference between Ascension Island and St. Paul's Rocks is extreme but may in a sense be due to a different level of sampling. The rocks of Ascension Island are representative of igneous activity which is taking place at a high level within the earth's crust, whereas at St Paul's Rocks we find dunites typical (we suppose) of a lower level.

Seychelles Islands

The Seychelles group of islands stand on a large shallow bank south and west of the Carlsberg Ridge in the Indian Ocean—this ridge being a part of the mid-ocean ridge system.[10, 55, 175, 178, 208] All except one of the islands in one of the groups which make up the Seychelles is composed of a hornblende-granite—an acid igneous rock containing the amphibole hornblende as a constituent; the age of this granite is 650 million years. It is cut by Precambrian and Tertiary medium- and fine-grained basic dykes. One island is composed of a Tertiary syenite and granite ring complex.

Seismic refraction measurements suggest that the whole of the Seychelles Bank is made of granite rock, and the ring complex is similar to those found in Africa, in Kenya and Nyasaland. These facts leave little doubt that the area is a remnant of a region which was once "continental", and could be called a micro-continent. It would be difficult to argue that the granites are only a local differentiate of local basalts, for they are so extensive. If the Seychelles were once a part of the African continent, or of Madagascar, some mechanism for the displacement is required. The suggestion has been made[175] that the trench which lies to the west of the Seychelles Bank—the Amirantes trench—

and the east coast of Madagascar represent the surface expression of a transcurrent or tear fault, a fault like those of the eastern Pacific Ocean floor in which there has been large horizontal displacement.

Hawaiian Islands

The Hawaiian Islands are a group of islands in the northern part of the central Pacific Ocean, and extend for about 2700 km in a west–north-west direction. They rest upon a submarine ridge 160 –320 km wide, 3200 km long, which itself rests upon a broad "swell" on the ocean floor. A deep, called the Hawaiian deep,[108] lies between the ridge and swell in the south-east part of the island chain, but is absent in the north-west part. The deep and the swell suggest some sort of failure of the crust to support the load of the islands. The islands to the west are older than those to the east[298] and the absence of a deep in the west may be due to filling in by sedimentary detritus.

The development of volcanoes sheds light upon the nature of the source of the lavas, and of the location of reservoirs of molten rock. The Hawaiian Islands are extensive, and have been studied fairly thoroughly, so that they may be considered as type examples of volcanic activity within the ocean basins. An evolutionary sequence has been recognized among the various islands, different islands giving evidence of the different stages in development. First, shield volcanoes with wide lateral extent are built by tholeiitic basalts from the sea floor. Second, the volcanic activity wanes and the central parts of the volcanoes collapse, presumably because empty chambers are left which formerly held lava. The lavas of this stage are principally alkali basalts and may contain fragments of gabbros and peridotites. The gabbro may represent rock which crystallized slowly within magma chambers themselves. The peridotite could be a source of the basic rocks, or result by accumulation of crystals of olivine which sank to the bottom of melts in magma chambers. Third, volcanic activity increases and the lavas are silica poor — nepheline bearing alkali basalts being common. Fragments of peridotites and eclogites are found.

The common feature to the sequence of lavas is the increase in alkalis (Na_2O and K_2O) so that the lavas are in turn tholeiites, alkali basalts and nepheline and melilite bearing alkali basalts. Two suggestions can be made to explain this, first that the alkali basalt is a later differentiate of the same source that yielded the tholeiite initially, and second that the lavas are generated within different regions of the mantle. The location of the sources of lava is not known with certainty, but the tracking of earthquakes associated with vulcanism has helped with this problem.

A particular study of vulcanism and earthquakes on Hawaii showed that before eruption the volcano swells and numerous tremors accompany the upward movement of lava.[64] After eruption the swelling subsides. Earthquakes associated with subsequent eruption occur at depths of 55 km. The conclusions drawn from the study were that lava was generated at these depths within the mantle and that the subsequent course of events was determined by the speed with which this magma was forced to the surface and spilled out as lava. If it was trapped for some time at depths of (say) 5 km, considerable differentiation and perhaps cooling could occur; if it was ejected quickly little differentiation could occur.

There are numerous studies of Hawaiian geology and petrology.[182−4,256,275,309]

The development of volcanoes of the Hawaiian Islands, as observed above sea-level, must not be taken to be representative of the development of volcanoes upon the floor of the ocean itself, represented topographically by many of the abyssal hills.[199] Ash forms in explosive volcanic eruptions through the expansion of gases within the lava. Their expansion is restricted at the bottom of the deep ocean because of the high hydrostatic pressure. Consequently volcanoes made substantially of ash are not likely to occur on the ocean floor. The high confining pressure will also tend to prevent cracks developing at the bases of volcanoes in deep water; this means that eruptions of lava low down on the flanks of the volcano will be uncommon. Consequently, voids will not be present and collapse structures such as those seen on the Hawaiian Islands will not form. Such arguments explain the absence of large collapse structures such as calderas on submarine volcanoes—only central vents are observed. The size of a volcano depends upon the possibility that the magma chamber can be incorporated within it as it increases in size. If not, and the chamber remains located well below the volcano's base, it will become extinct because the lava cannot be forced upwards sufficiently far. However, if the volcano is wide enough to allow the chamber to be incorporated within it, then it can remain active and continue to build upwards.

IGNEOUS ROCKS FROM THE MID-OCEAN RIDGES

Igneous rocks have been collected in dredge hauls from the Mid-Atlantic Ridge, the Carlsberg Ridge, the East Pacific Rise, and from seamounts and abyssal hills which rise from the Iberian abyssal plain.[69−71,191−2,195,209,217,229,341] The specimens from a number of these collections were quite fresh and little altered; those from the Iberian abyssal plain were, by contrast, badly altered but provide valuable insight into the course of weathering of igneous rocks on the ocean floor.

A variety of basic and ultrabasic rocks are represented in the collections—basalts, dolerites (diabases), gabbros and serpentinites among them. Basalts collected from the sides and the top of the east flank of the Mid-Atlantic Ridge at 45°50′N. included fragments of pillow lavas, with radial fractures which lead to their characteristic shapes, pyramidal with a curved base. The curved base may have a glassy margin, perhaps crystallized, up to 2 cm thick. A particular characteristic of these basalts is the presence of xenocrysts of magnesium-rich olivine, plagioclase very rich in calcium, and chrome spinel. The plagioclase was not in equilibrium with the liquid which finally crystallized. Muir and Tilley suggest that these characteristics arise because the basalts were derived from a layered gabbroic intrusion not far below the present level of the basalts. Such a layered intrusion can arise through settling of heavy minerals, such as olivine and spinel, during the course of crystallization. Fragments of the rocks which result could be incorporated as foreign elements in lavas.

The basalts collected on the sea floor are different from those typical of oceanic islands; chemical analyses of some of them are presented in Table 7.2. Those from the islands of the Atlantic Ocean are richer in titanium, potassium, sodium and iron, and poorer in silicon and calcium. Those from deep-water are tholeiitic, not alkali basalts, and are particularly poor in potassium and titanium by comparison with many basalts of islands and with younger basalts of continents.

Two reasons for these differences have been proposed, first that the alkali basalts are differentiation products of the tholeiitic basalts, and second that the alkali basalts are generated at deeper levels and higher pressures within the mantle. The second suggestion is in accordance with the observations that the alkali basalts are found on oceanic islands. These form an additional load upon the ocean floor, and will cause changes in pressure and temperature in the mantle beneath them. This may cause changes in the melting of eclogite. One interpretation of Yoder's and Tilley's observations is that garnet could be removed at high pressure from an eclogite, and this would lead to the generation of an alkali basalt. The pyroxene is richer in the normative minerals typical of the alkali basalt than the garnet. Conversely, removal of the pyroxene of the eclogite at low pressures would lead to a tholeiitic basalt.

The observations on basalts of the ocean basins suggest that the younger basalts of the continents are more contaminated by elements typical of the continental crust, such as potassium, than the basalts of the ocean floors. This is also suggested by problems which arise from the values of heat-flow over the surface of the earth, which demand that the mantles beneath continents be depleted in radioactive elements, such as rubidium and potassium. A corollary is that basalts now found on continents may not be "contaminated" if they were actually formed under deep-ocean conditions; examples of such basalts may be the Archaean Keewatin metamorphosed lavas of the Canadian Shield.[296] These rocks were originally volcanic, but now consist of metamorphic minerals. Wilson and others calculated the normative minerals from chemical analyses, and used the norms to suggest the original rock type.

They form a suite of rocks ranging in type from tholeiitic basalts to rhyolites—compare the chemical analyses 1 and 3 of Table 7.2. Igneous rocks found over the surface of the earth can be grouped into a number

Table 7.2. *Average composition of Canadian archaean lavas, oceanic tholeiites and other rocks*

	1	2	3	4	5	6	7
SiO_2	49·83	49·94	49·84	48·16	48·08	48·04	50·11
Al_2O_3	14·64	17·25	14·09	18·31	17·22	17·26	15·92
Fe_2O_3	3·03	2·01	3·06	4·24	1·32	1·19	1·58
FeO	8·77	6·90	8·61	5·89	8·44	8·56	7·80
MgO	7·36	7·28	8·52	4·87	8·62	10·24	8·67
CaO	10·46	11·86	10·41	8·79	11·38	11·26	11·30
Na_2O	2·02	2·76	2·15	4·05	2·37	2·38	2·63
K_2O	0·23	0·16	0·38	1·69	0·25	0·09	0·13
H_2O	1·18	—	—	—	1·06	0·34	0·66
CO_2	0·33	—	—	—	—	—	—
TiO_2	0·94	1·51	2·52	2·91	1·17	0·74	1·14
MnO	0·21	0·17	0·16	0·16	0·10	0·13	0·13
P_2O_5	0·19	0·16	0·26	0·93	0·16	0·09	0·12
	99·82	100·00	100·00	100·00	100·17	100·31	100·19

Notes to Table 7.2

1. Average composition of Canadian archaean lavas (53 samples): Wilson *et al*, Table 2.[296]
2. Average composition of oceanic tholeiite dredged from Atlantic and Pacific Oceans (10 samples): Engel *et al.*, Table 3 (water free).[71]
3. Average composition of tholeiites and olivine tholeiites from the Hawaiian Islands (181 samples): Macdonald and Katsura:[184] cited by Engel *et al.*, Table 3 (water free).[71]
4. Average composition of alkali basalt from submarine volcanoes and islands of the eastern Pacific Ocean (10 samples): Engel *et al.*, Table 3 (water free).[71]
5. Skaergaard intrusion: chilled olivine gabbro from marginal border group: specimen EG 4507: Wager.[287]
6. Average composition of two rock samples from the Mid-Atlantic Ridge: Nicholls *et al.* (1964), Table 2, col. 1.[217]
7. Average composition of three rock samples from the Mid-Atlantic Ridge: Nicholls *et al.* (1964), Table 2, col. 2.[217]

of associations reflecting chemical or mineralogical similarity, and there are numbers of quantitative tests for distinguishing the rocks of one association from another. Of these a well-known association is that of the tholeiitic flood basalts and intrusive quartz diabases,[279] which are approximately equivalent to rocks which have been called "calcic" or "calc-alkaline" by some. They are represented on the continents typically by the large floods of basaltic lavas of the Columbia and Snake River plains of the western United States and by those of the Deccan plateau in western India. The Keewatin lavas can be compared with this and other associations of igneous rocks, and it is found that they are similar to the tholeiitic basalts.

However, the same tests which lead to this conclusion lead also to the conclusion that the Keewatin lavas are similar to the tholeiitic basalts of the ocean floor. Perhaps the truth lies close to the suggestions of Engel *et al.*: "It is plausible, therefore, to relate the Archaean metatholeiites to their emplacement in a much thinner and less sialic continental nuclei than that invaded by younger continental tholeiites."

Basalts of the ocean basins are not always as distinctively tholeiitic or alkali-basalts as has been implied in the preceding paragraphs, and rocks of mineralogy and chemistry intermediate between the two extremes have been collected from the Indian Ocean and described by Cann and Vine.[317] Their nature may be seen by comparing the percentages of normative olivine, approximately 10·5%, and normative hypersthene, approximately 7·5%, with the values in Table 7.1 for a typical tholeiite and alkali basalt. These rocks were obtained in a dredge haul close to a fracture zone of the Carlsberg Ridge, and in dredge hauls close by meta-morphosed basalts were obtained which shed light upon the origin of sodium-rich lavas, spilites. Spilites are found on the continents in eroded geosynclines, and in time were associated with the later stages of the filling and sinking of geosynclines such as that of Devonian age in southwest England and Germany, and that of Eocene age in the western United States. They are basic lavas, which consist principally of sodium-rich plagioclase feldspar (not calcium-rich) and a pyroxene or its equivalent after alteration, such as actinolite. Their genesis has been a source of speculation — do they arise from a magma of the same composition of the lavas, or are they the product of alteration of basic lavas of more normal composition? Those described by Cann and Vine form a set which shows complete gradation from fresh basalt to spilite. The shape and arrange-ment of the mineral grains of the spilites are similar to those found in the basalts — the laths of plagioclase differ in composition, of course, and phenocrysts which were once olivine are now chlorite in the spilites. This transition is seen in the values for percentage of Na_2O found in chemical analysis, which ranged from 3·10 in the fresh basalt to 5·20 in the true spilite. Fresh, unaltered basalt is often collected from the ocean floor so that the mere extrusion of a hot lava into sea water does not lead to enrichment in sodium and spilitization; Cann and Vine suggest that the process arises after burial beneath 1 km of rock and with increase in temperature to approximately 200°C, by the action of hot fluids.

Other rocks in the haul from which the spilites came are more strongly metamorphosed. They represent ultrabasic material found between the pillows of the lavas, which have been altered to minerals such as chlorite, talc and actinolite. These minerals have in turn been replaced in increasing proportion by quartz; this is perhaps of great significance, because of the apparent rarity of quartz-rich rocks in the ocean basins.

Some comment should be made about the "andesite line". This line relates to the types of rock found around the margins of the Pacific Ocean and within it, and it is defined petrographically. Outside the line around the margins of the Pacific Ocean the volcanic rock association which predominates is the "andesite rhyolite basalt association" of volcanic orogenic regions. These are volcanic rocks with one major characteristic—they contain a large quantity of intermediate igneous rocks, andesites, by comparison with basalts. With this quantity of andesites a larger quantity of basalts would be anticipated than are found, and Turner and Verhoogen suggest that this phenomenon arises because of remelting of a variety of rocks in hot regions of unstable areas.[279] By contrast the rocks found inside the andesite line (the ocean side) are predominantly basalts, and the intermediate rocks associated with them occur in such quantities that there is no difficulty in supposing that they are differentiation products of the basalts. The andesite line coincides approximately with the margin of the Pacific Ocean defined topographically—with some of the island arc systems, for example.

BASALTS FROM THE IBERIA ABYSSAL PLAIN

We might suppose that if we are interested in igneous rocks which have been "contaminated" as little as possible between generation and eruption, volcanic rocks from the floors of the deep ocean basins would be most suitable for study. This may be so but the idea has to be tempered with caution, for they are relatively inaccessible, may be badly altered (see below) and further, if magma is generated at depths of 50–60 km in the mantle it could be argued that differences of 4–8 km (from the ocean bottom to sea-level and the tops of volcanic islands respectively) may make little difference to the degree of contamination. However, the point is probably that they will not be so contaminated by previous lavas and their related intrusives; certainly their study is worthwhile. What follows is an account of studies by D. H. Matthews[191–2] of rocks collected from the Iberian abyssal plain.

The Iberian abyssal plain lies in the eastern part of the North Atlantic Ocean. It is bounded by the continental slope off Spain and Portugal to the east and by abyssal hills to the west (Figs. 5.1, 5.4). It is connected at its north end to the Biscay plain by the gap in a range of hills called Theta Gap.[167] An account of this plain and its sediments can be found in Chapter 5.

The Geology of some Iberian Abyssal Hills

The Iberian abyssal plain lies in the eastern basin of the North Atlantic Ocean (Fig. 5.1). A group of hills lies in the western part of the plain and seamounts called the Western Seamounts lie west of the plain itself. The hills rise a few hundred metres and the seamounts about 2 km above the sea floor; their topography is rugged, not smooth, and they are associated with magnetic anomalies which run parallel to the hills (north–north-east trend), and run west–south-west over the seamounts. Matthews suggested that the seamounts and hills originated as volcanoes, first as fissures (which is the final stage reached by the Western Seamounts). Seismic refraction profiles suggested that the Mohorovičić discontinuity lies at a depth of about 10 km below sea-level.

The rocks which compose one of the hills, Swallow Bank, and the Western Seamounts are lavas. Those at Swallow Bank are vesicular basalts of tholeiitic affinity perhaps, and those at the Western Seamounts are vesicular basalts associated with oligoclase-basalts: their affinity may be with alkali-basalts. (Oligoclase-basalts are in a formal, terminological sense not truly basalts, which typically have plagioclase feldspar in which the percentage of the pure calcium end member of the plagioclase feldspars $CaAl_2Si_2O_8$ is greater than 50%.) The rocks are fragmented and badly weathered and coated on the outside with a manganese–oxide type of coating; this may often cement shattered fragments together and serves, together with the petrological homogeneity, to indicate that the collection of dredged rocks is not one of erratics, but represents the rocks in place. The lava was extruded as massive flows and sometimes as pillow lavas.

The weathering was complicated and has been divided by Matthews into four stages.

(1) *Oxidation and hydration*

The glassy margin (sideromelane) caused by chilling on extrusion of the lava into water is devitrified (loses glassy nature) and hydrated: the product is called *palagonite*. The interstitial glass of the lava is oxidized and hydrated to *chlorophaeite*, i.e.

glassy selvage (sideromelane)———→palagonite
glassy mesostasis ——→ chlorophaeite.

(2) *The formation of zeolites and feldspars*

(a) Calcium-rich plagioclase altered to potassium-rich feldspar (adularia) or harmotome (a barium rich zeolite) or both together.

(b) Zeolites and calcite are formed within the chlorophaeite matrix of stage (1).

(3) *Clay mineral formation (argilization)*

(a) The yellow palagonite of stage (1) altered to an aggregate of montmorillonoids (clay mineral) and the cryptocrystalline silica, chalcedony.

(b) The chlorophaeite of stage (1) altered to an aggregate of montmorillonite (clay mineral often associated with volcanic rocks in the sea) and chlorites (micaceous mineral), with iron oxides.

(4) *Cold weathering*

(a) Yellow palagonite of stage (1) altered to montmorillonite.

(b) Radioactive pelagite (manganese oxide mainly) deposited in cracks and on exposed surfaces.

In summary, a high-temperature assemblage of minerals has been weathered to a low-temperature assemblage. Montmorillonite is well-known as accompanying submarine volcanoes (see Chapter 4), and the weathering processes suggested by Matthews explain this feature of its occurrence. The radioactivity of the weathered lavas is surprisingly high. In terms of equivalent heat production it is about the same as a "typical" granite, and the manganese crust on the lavas is yet more radioactive. This high radioactivity in the altered lavas may be due to the pelagite deposited in stage (4), and can have been acquired only from the sea water. The original lava (of which little remained) is not especially radioactive.

The seismic velocities of compressional waves in the altered lavas are low, in the range 3·0–3·5 km/s; the range in the most fresh lavas is 3·5–5·5 km/s. This is of course most valuable information for it helps us to interpret results of seismic exploration of the sea floor more easily. In Chapter 8 we will see that there is good evidence that rocks exist in the oceanic crust which have seismic velocities (compressional waves) in the range 4·0–6·0 km/s, and results such as these are good evidence that such rocks are made of lavas and sediments, deposited one upon another.

Lavas from the deep ocean floor, such as these are, are of especial value because they have been poured out at a level only 4 or 5 km above the Mohorovičić discontinuity. One might think that the level at which they originated could be estimated from the magnetic anomalies associated with the abyssal hills and seamounts. Matthews points out that this cannot be done, for the susceptibility contrast between a gabbro (of which the lower part of the oceanic crust might be made) and the postulated feeder dyke of the lavas may be too small. Nevertheless, the presumption is that these lavas came from a source within the mantle: the temperature gradient would have to be especially high for melting to occur above the Mohorovičić discontinuity. Yoder and Tilley attempted to deduce the probable course of vulcanism in depths of water from which Matthew's rocks were dredged and it is interesting that the conclusions they reached are similar to his observations.

MANGANESE IN THE OCEAN BASINS

Manganese oxides are intimately associated with the volcanic rocks of the Iberian abyssal plain; any general discussion of the origin of these oxides must be related to the distribution of manganese in the ocean basins as a whole. Manganese is not an abundant element in the earth, either within the crust, or within the interior of the earth if we may take chondritic meteorites as a guide. Its abundance in average igneous rocks is approximately 0·1%, and that of iron for comparison is several per cent. Values for the abundance of manganese are shown in Table 7.3. The relatively low values are quite evident and interest in manganese arises because of the presence on the ocean floor in some parts of the world of slabs and nodules of manganese-rich material, in which the concentrations of manganese and iron are approximately the same, in the ranges 11–24% and 6–20% respectively.[234]

The sources of supply of the manganese found in oceanic sediments are ultimately the igneous rocks of continents and ocean basins. Manganese does not in general form separate manganese-dominated minerals, but as the ion Mn^{2+} can substitute for Fe^{2+}, Mg^{2+} and Ca^{2+} in a variety of minerals of igneous rocks in which the other ions are in six-fold coordination. The ionic radii of these ions are shown in Table 7.5, and the close similarity of Mn^{2+} and Fe^{2+} in both size and charge and other properties[325] makes the similarity in behaviour of the two ions likely.

Table 7.3. *Abundance of manganese*

Igneous rocks	
1. Ultrabasic	1620 parts per million
2. Basaltic	1500
3. Syenites	850
4. Granites	465
5. Oceanic tholeiites	1240
6. Alkali basalts	1160
Sedimentary rocks	
7. Shales	850
8. Sandstones	0
9. Carbonates	1100
Deep-sea sediments	
10. Carbonates	1000
11. Clay	6700
Sea water	
12. Mn	0·0007–0·016
13. Mn^{2+}	0·0002–0·003
Particulate organic matter in sea water	
14. Mn^{2+}: shallow samples	0·00072
15. Mn^{2+}: deep samples	0·000038

Note: 1–4, 7–11 from Turekian and Wedepohl.[278]
　　　 5,6　　　　 from Engel and Engel.[70]
　　　 12　　　　　from Richards.[333]
　　　 13　　　　　from Wangersky[337] citing Rona.
　　　 14,15　　　 from Wangersky[337]

Table 7.4. *Chemical composition of manganese nodule (from Goldberg[323])*

Element	Range %
Fe	6·2–20
Mn	11–24
Ni	0·16–1·1
Co	0·01–0·7
Cu	0·17–1·8
Th	0·0020–0·0130
Pb	0·0500–0·2500

Table 7.5. *Ionic radius and coordination numbers* [335]

Ion	Mn^{2+}	Mn^{3+}	Mn^{4+}	Mg^{2+}	Fe^{2+}	Fe^{3+}	Ca^{2+}
Radius (Ångstroms)	0·80	0·66	0·57	0·66	0·74	0·64	0·99
Coordination	6	6	4	6	6	6	6

Consequently, manganese as Mn^{2+} is most abundant in basic igneous rocks with a high content of ferromagnesian minerals (olivines and pyroxenes) and least abundant in SiO_2-rich igneous rocks such as granites. Both Mn^{2+} and Fe^{2+} can be leached out of igneous rocks by acidic and reducing solutions and reprecipitated as MnO_2 and Fe_2O_3 (Mn^{4+}, Fe^{3+}) under oxidizing conditions. Hence a high proportion of the manganese leached out of rocks on the continents will not reach the oceans; what does reach the open oceans does so as parts of the detrital mineral grains carried there, or in solution. The amount of manganese in solution is small—a few parts in 10^9, and its speciation is not certain,[253] but some is as Mn^{2+}.[337] As Mn^{2+} in solution it is apparently under-saturated by two orders of magnitude;[323] that this is puzzling is seen if a residence time t is defined as

$$t = \frac{\text{amount of element in the sea in grams}}{\text{amount sedimented per year}}$$

The value of t is approximately 7000 years.[337] It would be better, in view of the uncertainty of the value of the amount sedimented per year, to know the residence time defined by Barth:[12]

$$t = \frac{\text{amount of element in the sea in grams}}{\text{amount supplied per year}}$$

The value of t of 7000 years suggests that some mechanism is operating which extracts manganese from sea water and transports it to the bottom, in spite of the low concentration in sea water. This can take place through organic agencies. Arrhenius has reported that a collection of planktonic Foraminifera contain 8 ppm manganese.[7] Wangersky and Gordon have found several parts in 10^9 of manganese as Mn^{2+} in particulate organic matter from the oceans.[337] The Foraminifera and the particulate matter will sink to the sea floor in time. Manganese liberated close to the sea floor is oxidized and precipitated as MnO_2. Several observations have been made that Manganese-rich bands are found in the upper parts of sediment cores.[315,323] The mechanism for this enrichment is that manganese is transferred at depth within the sediment into a relatively soluble fraction, and with expulsion of interstitial water on compaction moves vertically towards the oxidizing conditions which prevail near the sediment-water interface, where it is precipitated. Elements other than manganese are also extracted from sea water and transported to sediments by biological agencies.

The contribution of manganese from the igneous rocks of the ocean basins themselves (rather than that derived by weathering of continental rocks) is associated with the problem of the origin of manganese nodules.

The term "nodules" has been used as a catch-all name for a variety of types of occurrence of manganese. Typically parts of the ocean floor are covered to the extent of 20–30% areally by slabs of manganese-encrusted material tens of centimetres in dimensions. Often the manganese is associated with volcanic rocks either as an encrustation cementing fragments of lava together, or very intimately in a way which demonstrates that the manganese-rich material was derived from the lava itself when it was extruded into sea water. Matthews,[191−2] and Bonatti and Nayudu[315] have described such occurrences in detail. A nodule from the Mendocino Ridge in the Pacific Ocean for example possessed a dark-brown crust of manganese and iron oxides through which protruded basalt fragments. In the interior of the nodule palagonite was found—an hydrated devitrified glass which originates from sideromelane, the glass material which is generated by rapid chilling on extrusion of a lava into water. This palagonite enclosed fragments of the sideromelane. The association of manganese oxides with volcanic rocks has been well documented; what is not explained quantitatively is the concentration of manganese from the lava. Bonatti and Nayudu suggest that the sea water provides a good solvent for acidic gases present in the lava, which leads to leaching of the manganese and iron. These elements are precipitated as oxides as the acid is neutralized by sea water, and as the environment becomes more oxidizing in the presence of fresh sea water. Note that this implies a rapid formation of manganese nodules, not a slow one: mechanisms which suggest slow accumulation of manganese-rich oxides are difficult to accept because the rates of accumulation postulated are sometimes slower than the rate of accumulation of the local sediment, yet the nodules are not buried. Studies of the details of the mineralogy and geochemistry of the manganese–iron oxides support the mechanism proposed by Bonatti and Nayudu.[315,316] The oxide formed at low oxidation potential is a disordered mixed layer structure consisting of alternating sheets of MnO_2 and $Mn(OH)_2$—that is, not all the Mn^{2+} is oxidized. In this disordered structure it is easy to substitute ions of different sizes, such as cobalt. Such a structure would form under conditions of rapid formation, in which the concentration of Mn^{2+} is high and of oxygen in the water relatively low. The oxides formed at higher oxidation potentials are single sheet structures of MnO_2—all Mn^{2+} oxidized; this structure is ordered, and foreign ions can only substitute with difficulty. Higher oxidation potentials would be characteristic of slow precipitation of manganese from dilute solution in sea water, with a normal oxygen content. Thus both the order–disorder relationship and the content of foreign ions such as cobalt should reflect the mode of formation.

Arrhenius and others find that the ratio cobalt/manganese is higher in nodules further from the continental margins. Consequently, they suggest that those near the continents may be formed by slow precipitation from the manganese in solution in sea water, whereas those further away from the margins originate in volcanic processes in rapid reactions.

The origin of manganese nodules has been puzzling for many years and numbers of solutions to the puzzle have been suggested of which that of Matthews, Bonatti and Nayudu is only one. However, it seems the most plausible at the present time, and the situation now is probably that enough is known for definitive experiments to be undertaken.

INHOMOGENEITIES WITHIN THE UPPER MANTLE

Isotopic processes

Evidence is accumulating from seismic studies that the upper part of the mantle of the earth is not homogeneous, and this will be discussed in Chapter 8. Similarly, evidence is accumulating from studies of isotopes of rubidium and strontium, and of uranium, thorium and lead that the mantle is inhomogeneous.[93,94,226,248] The study of isotopes of these elements is useful not only because of the information about the mantle, and the ages of rocks thereby gained, but also because they lead to an understanding of some geological processes.[113,206] For example, a study by Moorbath and Bell led to the conclusion that granites intruded high up within the earth's crust were generated by fusion of ancient crust, not by differentiation of a basic magma within the mantle.

Rubidium-87 decays radioactively to strontium-87 (see Chapter 3). Suppose a rock crystallizes from a melt at time t years ago and is collected by us now. Then the Sr^{87} content measured now depends upon the initial content of strontium-87 and the initial content of Rb^{87}. Since the time of crystallization t years ago, if the system has remained isolated with respect to rubidium and strontium, then the strontium content has changed only by addition of radiogenic Sr^{87} generated by the decay of Rb^{87}. In fact, expressed as ratios with respect to non-radiogenic Sr^{86}, we can write for one mineral in the rock,

$$(Sr^{87}/Sr^{86})\text{ today} = (Sr^{87}/Sr^{86})\text{ initial} + (Rb^{87}/Sr^{86})\text{ today} \ (e^{\lambda t}-1)$$

$$\text{mineral 1} \qquad\qquad \text{mineral 1} \qquad\qquad \text{mineral 1}$$

and for a second mineral,

$$(Sr^{87}/Sr^{86}) \text{ today} = (Sr^{87}/Sr^{86}) \text{ initial} + (Rb^{87}/Sr^{86}) \text{ today} \quad (e^{\lambda t} - 1).$$

   mineral 2   mineral 2   mineral 2

The quantities (Sr^{87}/Sr^{86}) (today), and (Rb^{87}/Sr^{86}) (today) for both minerals can be measured, from which the two unknowns t and (Sr^{87}/Sr^{86}) (initial) can be found, provided that this initial ratio was the same for both minerals.

It is found that the ratio (Sr^{87}/Sr^{86}) (initial) is relatively low in recent oceanic basalts, and rather uniform by comparison with the values found in other rock types. For example, its value is approximately 0·705 for recent basalts, but the value in the Tertiary granites of Skye is 0·712. Moorbath and Bell suggest that this means the granite was not the end product of differentiation of basic rocks associated with the granites, in which the value is 0·706.

The variation among basalts themselves is quite small. The initial Sr^{87}/Sr^{86} ratios in a number of silica-poor rocks from Oahu and Hawaii range from 0·7030 for a melilite–nepheline basalt of SiO_2 content 36·3%, to 0·7043 for a trachyte of SiO_2 content 62·0%. The volcanic rocks are so young that very little radiogenic Sr^{87} has been generated since emplacement, so that the variations reflect variations in the source, that is, within the mantle. Similar conclusions were reached in a study of volcanic rocks from Ascension Island and Gough Island in the Atlantic Ocean.

Fractionation of Rare-earth Elements

Fractionation of an element in the development of suites of igneous rocks is seen most obviously in the distribution of silicon — granites are enriched and gabbros depleted in the element. One group of elements which proves to be useful in studying fractionation among igneous rocks are the rare-earth elements, the fifteen in the lanthanide series from atomic number 57 (lanthanum) to 71 (lutetium).[327,335] The radius of the atoms decreases with increase in atomic number (Table 7.6) so that lanthanum is the largest (1·14Å radius) and lutetium the smallest (0·85 Å radius), and the size, rather than any other property governs their behaviour in melts and consequently their distribution. Most form tri-valent cations, except cerium which can be oxidized to Ce^{4+} and europium which can be reduced to Eu^{2+}. They tend to form element–oxygen bonds which are covalent by comparison with the ionic calcium–oxygen bond,

Table 7.6. *Rare-earth elements*

Element		Atomic number	Radius (Å)
Lanthanum	La	57	1·14
Cerium	Ce	58	1·07
Praseodymium	Pr	59	1·06
Neodymium	Nd	60	1·04
Promethium	Pm	61	
Samarium	Sm	62	1·00
Europium	Eu	63	0·98
Gadolinium	Gd	64	0·97
Terbium	Tb	65	0·93
Dysprosium	Dy	66	0·92
Holmium	Ho	67	0·91
Erbium	Er	68	0·89
Thulium	Tm	69	0·87
Ytterbium	Yb	70	0·86
Lutetium	Lu	71	0·85

Note: radius from Taylor.[335]

Table 7.7. *Abundance of rare-earth elements*

Rock type	Abundance (ppm)
1. Chondritic meteorites	5
2. Peridotite from St. Paul's Rock	32
3. Peridotites from Lizard, Cornwall; Mt. Albert, Quebec; Tinaquillo, Venezuela	1·07–3·88
4. Oceanic tholeiites	74–126
5. Gough island olivine basalt	280
6. Gough island trachyte	1037
7. Acid and intermediate rocks (SiO₂ 60–70%)	315
8. North American shales	230

Note: data from Haskin and Frey.[327]

and therefore as a group tend to concentrate in residual melts. Calcium is the most common element closest to the rare-earth elements in size (Table 7.5). The smaller (and heavier) rare-earths are enriched in the earlier fractions of crystallization, and the larger (and lighter) in the later fractions.

Consequently, ferromagnesian minerals such as pyroxenes and hornblendes are enriched in the smaller, heavier rare-earth elements, and feldspars enriched in the larger, lighter elements. Although high

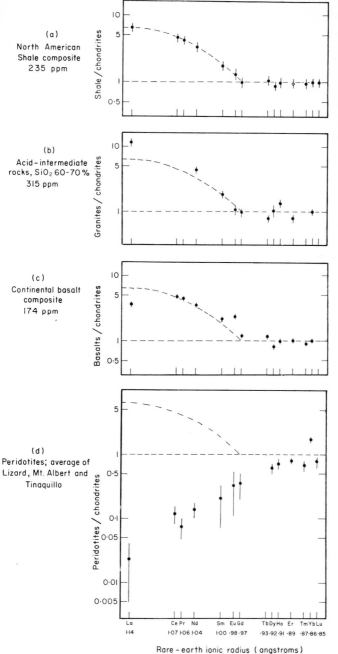

Fig. 7.2. Comparison plot of the rare-earth content of a number of rocks. The absolute abundances of each element have been divided by that of ytterbium, and then by the corresponding average values for twenty chondritic meteorites. The number by each is the total rare-earth content in parts per million. (a) Composite of forty North American shales. (b) Composite of granites and intermediate rocks, SiO_2 per cent 60–70. (c) Composite of continental basalts. (d) Ultrabasic rocks: average of Lizard, Cornwall; Mt. Albert, Quebec; Tinaquillo, Venezuela. The curve represents the values for North American shales (after L. A. Haskin and F. A. Frey[327]).

concentrations are to be found in accessory minerals such as apatite, these are quantitatively less important, and the abundance in igneous rocks is dominated by their concentration throughout the major rock-forming minerals. The gross features of the distribution of the rare-earth elements is seen in Table 7.7; they are absolutely less abundant in the most basic rocks (silica poor) such as chondrites and peridotites, and more abundant in the acid and intermediate rocks such as granites and trachytes.

The abundances of individual rare-earth elements alternate from high to low values as the atomic numbers are even and odd; for a composite of forty North American shales analysed by Haskin and Frey[327] the values for lanthanum, cerium, praseodymium and neodymium are 39, 76, 10·3 and 37 ppm, for example. Haskin and Frey have shown how revealing it is to compare abundances with a relevant standard and plot against atomic number or against ionic radius. Typical plots are illus-trated by Figs. 7.2 and 7.3. The absolute abundance of each element in any one rock type is first normalized by dividing by the ytterbium content, and then divided by the corresponding value of the standard. In the cases illustrated the standard has been taken as the average of twenty chondritic meteorites because this material may be representative of the solar system. Consequently the plots illustrate relative enrichment or depletion in the rare-earth elements.

A number of rocks show relative enrichment in the light elements compared with the chondritic pattern (and absolute enrichment overall by comparison with chondrites). These include sediments from North America, represented by a composite made of forty different shales, acid and intermediate igneous rocks (such as granites and diorites) and continental basalts; these are shown as the upper three plots of Fig. 7.2. The sediments' enrichment in light elements reflects their sources — the igneous rocks from which they were derived. If the pattern of the mantle and crust overall is to be similar to that of chondrites there should be found a rock deficient in the light elements. Such a rock may be represented by some peridotites, such as Mt. Albert peridotite of Quebec; values for a composite of three is shown as the bottom plot of Fig. 7.2. This implies that such peridotites could be representative of the upper mantle, but not the whole mantle if the absolute abundances are to be comparable with those of chondrites. Consequently lower in the mantle the content of light rare-earth elements must increase towards the chondritic values. This hypothesis rests upon the assumptions that the sediment, granite and continental basalt pattern is represent-ative of the continental crust, that the whole-earth pattern is like that of chondrites and that the Alpine peridotites such as Mt. Albert represent

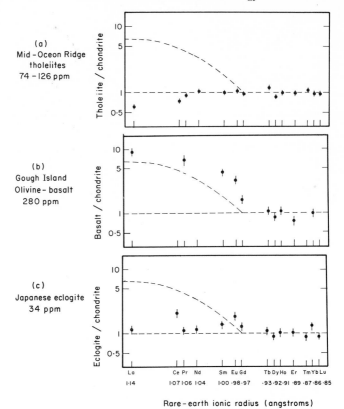

Fig. 7.3. Comparison plot of the rare-earth content of (a) tholeiites: average of seven from mid-ocean ridges. (b) Gough Island
olivine basalt. (c) Eclogite from Japan. See Fig. 6.12 caption (after
L. A. Haskin and F. A. Frey[327]).

the upper mantle. A number of observations by Haskin and Frey point
to alternative hypotheses.

The absolute abundances in gabbros are higher than in chondrites
and their distribution pattern is not like that of basalts, but more similar
to the eclogite pattern of Fig. 7.3. (bottom). Further, other ultrabasic
rocks are known in which there is enrichment in the light elements,
comparable with that of the North American shales, for example.
Tholeiitic basalts from mid-ocean ridges have a pattern similar to that
of chondrites, but deficient in lanthanum and cerium (Fig. 7.3). By
contrast, alkali basalts such as those of oceanic islands — Gough Island
is illustrated — are enriched in the light elements, and the more acidic
lavas of the alkali basalt series are overall greatly enriched in the rare-

earth elements. The pattern in one eclogite is comparable to that of the chondrites.

The distribution of the rare-earth elements between the garnet and pyroxene phase of the eclogite of Fig. 7.3 may be most relevant. The light elements are concentrated in the pyroxene, the heavy elements in the garnet; consequently if a tholeiite were generated by subtraction of a substantial part of the pyroxene and the alkali basalt by subtraction of garnet, distributions similar to those observed would result. The importance of tholeiites relative to alkali basalts is illustrated by calculating the relative volumes of each required to yield an overall chondritic pattern—90% of the mixture would be classed as tholeiitic from its composition of major elements.

SUMMARY

The main points which emerge concerning our present knowledge of the igneous rocks of the ocean basins are these:

(1) Acidic rocks are rare among oceanic igneous rocks. Small quantities are found on a few islands such as Ascension Island. One micro-continent known in which acidic rocks predominate is the Seychelles Bank.

(2) Basalts are the most commonly found igneous rock. The basalts of the oceanic islands are in part tholeiitic, in part alkali-basalts. The majority of the basalts of the deep ocean floor which have been studied so far are tholeiitic, and these are relatively uncontaminated by elements such as potassium which might be considered typical of the crust of the continents. Studies along these lines may help answer questions concerning the origin of basalts found now upon the continents.

(3) Ultrabasic rocks are not uncommon; they are found as peridotite and dunite and altered as serpentinite. The peridotite and dunite could be representative of the mantle beneath the ocean basin, or of one extreme product of differentiation which produced basalts as the other product. The altered ultrabasic rocks, serpentinites, must be formed at relatively low temperatures —a few hundred degrees Centigrade at most. This matter is considered in Chapter 8.

CHAPTER 8

The Structure
of the Ocean Basins

It will be clear by now that the geology of the ocean basins is different
from that of the continents. The structure of the crust and the upper
mantle beneath the oceans is also different from that beneath continents;
there are differences too between various parts of the ocean basins.
These differences have been found by means of geophysical measurements described in Chapter 2. It may be hard to relate the results of
physical measurements to description in terms of rocks of specified
chemical and mineralogical composition. Often it may be better to
describe a structure correctly as changes in physical parameters until
the relationship of these parameters to rock types is more surely known.
A combination of measurements of several different physical parameters
may, of course, lead to one rock being preferred above all others. It
seems likely that many of the principal features of the structure of
the oceanic crust have been found. However, the characteristics of the
upper mantle are certainly not well known and, until they are, syntheses
which attempt to explain the major features of the ocean basins (and
indeed the whole earth) will be unsatisfactory.

THE MID-OCEAN RIDGE SYSTEM

The mid-ocean ridges extend throughout the world's oceans, and
appear in places to continue beneath continents. The crust beneath
their crestal provinces is different from that elsewhere, but they are
in gravitational equilibrium with other parts of the ocean basins. A
large magnetic anomaly is often found along their crests, and the crests
themselves may be offset by fracture zones. Shallow earthquakes are
found beneath the crests, and particularly beneath fracture zones.
The rate at which heat flows out of the ocean floor is often higher
beneath the central parts of the ridges than elsewhere. Islands which
rise from them may be intensely volcanic.

The geometry of the ridge system is not particularly simple. Some of the principal members, the Mid-Atlantic Ridge, for example, are a median line between continents. Others, such as the East Pacific Rise and the ill-known Galapagos–Chile Rise[202] are certainly not median lines. Some appear to lie on the arcs of circles centred on the continental shields. Some continue beneath continents; the Carlsberg Ridge enters the Gulf of Aden and appears to continue as the Red Sea on one hand, as the East African Rift valleys on the other (Figs. 3.13 and 3.14). The features of the continents in such regions are similar to those of the ridges in some ways: they are elevated, fractured into ridge and trough provinces, associated with vulcanism—as in East Africa, or in western North America (the Columbia River basalts) associated with shallow earthquakes[263,265] and with rift valleys similar to the central valley of some of the mid-ocean ridges. Consequently one is led to propose that regions such as the Red Sea are embryo ocean basins, the continental mass beginning to split apart. The upper part of the mantle beneath oceans should be different from that beneath continents, for reasons associated with heat-flow and the earth's gravity field, discussed later in this chapter. However, mid-ocean ridges are topographically large features and if the East Pacific Rise continues beneath western North America its eastern flank will occupy a substantial fraction of the area of that part of the continent. Consequently differences between various parts of the upper mantle may not be controlled only by the presence or absence of a continent, but by the presence or absence of a "mid-ocean" rise. (The term "mid-ocean" being now, of course, a curious one to use.) This would lead to thermal disturbances in the parts of continents which overlie the high thermal anomalies of the crests of mid-ocean rises, and this in turn would lead to crustal instability and vulcanism.

The characteristic rough topography of mid-ocean ridges, and the central valley of some of them (Figs. 3.15 and 8.1), will be controlled at least in part by faulting, parallel to the general trend of the particular ridge. The faulting could arise in many ways, but an upwards push from beneath distributed along a line would lead to linear fractures. The central rift valley itself could be a collapse structure, in the sense that calderas are collapse structures caused by the presence of empty spaces vacated by lava-flows, or could be caused by a horizontal tension or pulling-apart centred at the crest of the ridge. Drag beneath the crust in the appropriate directions would be caused by convection currents rising in the mantle beneath the crests of ridges. Alternatively, intrusion of dykes in profusion, which has led to the spreading of Iceland (Chapter 7), would have a similar effect. If a principal focus

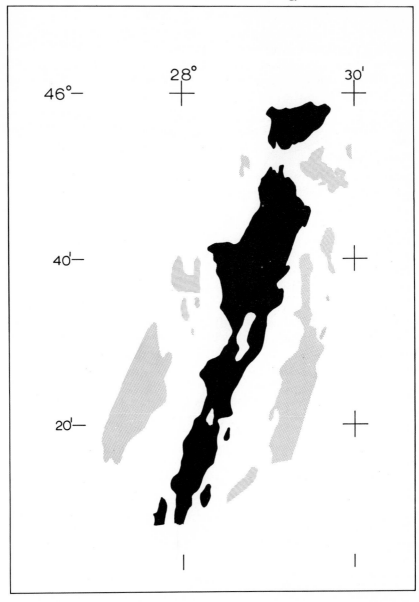

Fig. 8.1. The central valley of the Mid-Atlantic Ridge. Note the continuity of
the valley for at least 70 km and a blockage at the northern end. Solid black,
below 1500 fm; stipple, above 1000 fm. Based on surveys by C.S.S. Hudson and
R.R.S. Discovery II, with sounding lines predominantly east–west and approximately 2 km apart. Navigation based on moored radar-transponding buoys
(*HUDSON*) and radar-reflecting buoys (*DISCOVERY II*) (after B. D. Loncarevic[177]).

of volcanic activity is beneath mid-ocean ridges the centre of pushing apart by dykes would lie beneath the crests of ridges. The lateral spreading would vary with depth—presumably more dykes do not reach the surface than do. Below the level of generation of magma, which from the studies of earthquakes and vulcanism on Hawaii can be set at several tens of kilometres there would be fewer dykes. We might expect to find a slippage of the crust and upper mantle away from ridge crests.

The topographic features of the various parts of the ridge system are not all alike. For example, there is no median valley along the crest of the East Pacific Rise. This has led Menard[201] to emphasize the different stages of development which have been reached by the ridges in different regions—the East Pacific Rise being the youngest, the now defunct Darwin Rise (Chapter 7) the oldest for which we have evidence. Perhaps there has been system upon system of mid-ocean ridges over the last few hundreds of millions of years (at least).

The structure of the normal oceanic crust is described in Table 8.1 at the beginning of the next section. Several features should be noticed: the major part of the crust is 4–5 km thick and is associated with a compressional wave velocity of $6 \cdot 7$ km/s; the overall crustal thickness is near to 10 km; and the velocity of compressional waves in the mantle beneath is $8 \cdot 1$ km/s. The structure beneath the central parts of mid-ocean ridges is different. Seismic refraction studies and the measurement of gravity suggest that the Mid-Atlantic Ridge is underlain by 3–5 km of rocks in which the compressional wave velocities range from $4 \cdot 5$ to $5 \cdot 5$ km/s, which correspond to those which would be expected in volcanic rocks (and also in consolidated sedimentary rocks), and beneath this "layer" is a zone approximately 35–40 km thick in which the compressional wave velocity is about $7 \cdot 3$ km/s.[83,168,269] This velocity is intermediate between that which is commonly found in the upper mantle in seismic refraction studies ($8 \cdot 1$ km/s) and that typical of the oceanic crust itself ($6 \cdot 7$ km/s); Ewing and Landisman[83] make the interesting observation that it may be that the velocity does increase to $8 \cdot 1$ km/s at some depth (say 40 km), but that the values of $7 \cdot 3$–$7 \cdot 5$ for the velocity arise either through extension of mantle material near to the surface of the earth, the velocity decrease being caused by the change in pressure, or through mixing normal oceanic crustal rocks with mantle rocks. Similar velocities of propagation of compressional waves have been reported from the East Pacific Rise.[198] The higher temperature gradient beneath the crests of ridges will also lead to a decrease in the compressional wave velocities in the hotter rock, and this effect, although it has been considered, may not have had the attention it deserves.

The East Pacific Rise which differs topographically from the Mid-Atlantic Ridge, has a different crustal structure, perhaps reflecting also a difference in stage of development. The main part of the crust (layer 3 of later pages) thins from near 5 km in other parts of the Pacific Ocean to near 4 km under the crestal regions. As beneath the Mid-Atlantic Ridge the velocities beneath the crest are abnormally low. The low velocities found in the upper part of the mantle can be due to crust-mantle mixing, and to high temperatures. If they are due to a mixing process the rocks which result will be lower in density than normal mantle rocks, and so will compensate for the elevated topography. Gravity observations assist us here.

Measurements of g throughout the oceans have been made by pendulum measurements[338] and surface-ship gravimeters.[173] The free-air anomaly over the Mid-Atlantic Ridge is near zero: this can be seen from Fig. 8.5. This implies that the major features of the topographic elevation of the ridge, which so to speak displace lighter water, must be accounted for at depth by rock less dense than usual. The detailed topographic features are not compensated, as we might expect, the profiles of the free-air anomaly resembling in shape the smoothed bathymetry. Consequently it is reasonable to postulate that the rocks down to a few tens of kilometres beneath the Mid-Atlantic Ridge are less dense than is normal in the mantle. Detailed studies of the gravity field over mid-ocean ridges have not been made so far over a sufficently large area for sound conclusions to be drawn, but the first part of such a detailed study is shown in Fig. 8.2. This shows the Bouguer anomalies, derived from the free-air anomalies by adding the effect of rock replacing the water.[176] If the density of the rock is chosen to be equal to the density of the crust down to a particular level the resulting anomaly map will reflect anomalous mass distributions below that depth. The difficulty which has to be faced is the choice of density. This can be estimated from the compressional wave velocity. In the particular example shown the density of rock replacing water was chosen to be 2·67 g/cm³. We see that there are variations in the Bouguer anomalies of approximately 20 mgal. These could arise through a combination of errors in measurement and lack of knowledge of the bathymetry, but it is most likely that the variations represent the effect of density variations in the rocks beneath. If the density chosen is too low the anomaly map resulting will still resemble the topography; if the density is too high the map will resemble the upside-down topography. In this case the variations in Bouguer anomaly can be minimized, but not eliminated.

The crest of the Mid-Atlantic Ridge is associated with a large magnetic anomaly[130,157,268] (Figs. 8.5, 8.6, 8.7 and 8.8). This has a magnitude of

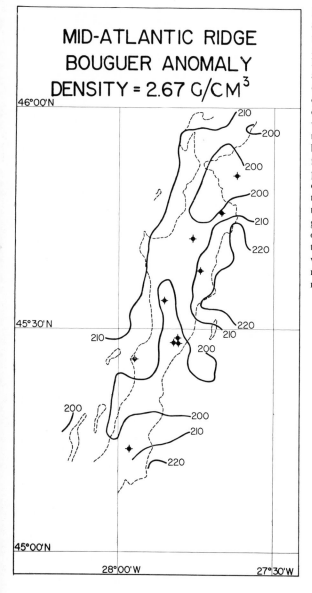

MID-ATLANTIC RIDGE
BOUGUER ANOMALY
DENSITY = 2.67 G/CM^3

Fig. 8.2. Bouguer anomaly over the central valley of the crest of the Mid-Atlantic Ridge. Heavy lines are anomaly contours at 10-mgal intervals, dashed lines are 1400-fm contours delineating the central valley. These gravity anomalies result after the effect of the water has been removed from the free-air anomalies by replacing it with rock of density 2·67 g/cm^3. Note there is no correlation between anomaly and topography; although the size of anomalies is sensitive to the density chosen, variations of Bouguer anomaly can be minimized but not eliminated (after B. D. Loncarevic[176]).

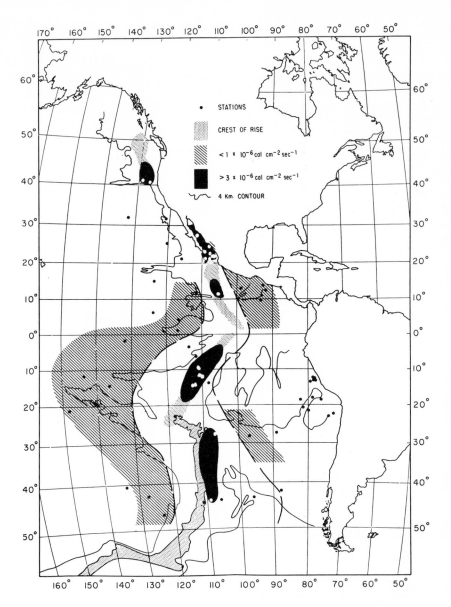

Fig. 8.3. Heat flow over the East Pacific Rise in the eastern part of the Pacific
Ocean (after H. W. Menard[198]).

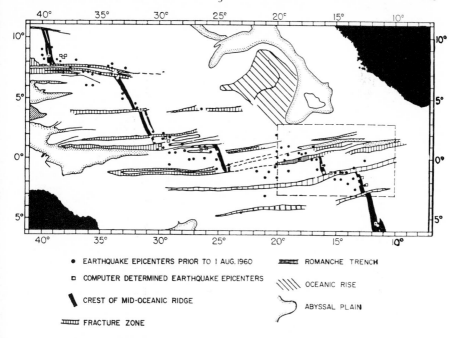

EARTHQUAKE EPICENTERS PRIOR TO 1 AUG.1960   ROMANCHE TRENCH

COMPUTER DETERMINED EARTHQUAKE EPICENTERS

OCEANIC RISE

CREST OF MID-OCEANIC RIDGE

ABYSSAL PLAIN

FRACTURE ZONE

Fig. 8.4. Physiographic features and earthquake epicentres in the equatorial Atlantic. Physiographic features after Heezen and Tharp.[128] Epicentres from Gutenberg and Richter[105] and U.S. Coast and Geodetic Survey epicentre cards. The dashed rectangle shows the limits of a detailed survey (after B. C. Heezen, E. T. Bunce, J. B. Hersey and M. Tharp [116]).

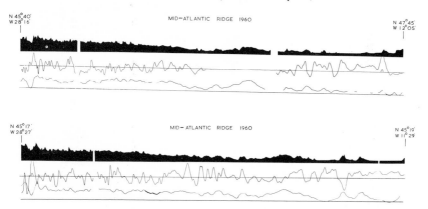

Fig. 8.5. The Mid-Atlantic Ridge: the two profiles show bathymetry (solid black), anomalies in total intensity of magnetic field (middle trace) and free-air gravity anomalies (bottom trace). Note the change in character of the magnetic field towards the flanks of the ridge—the dominant wavelengths increase (after B. D. Loncarevic, unpublished).

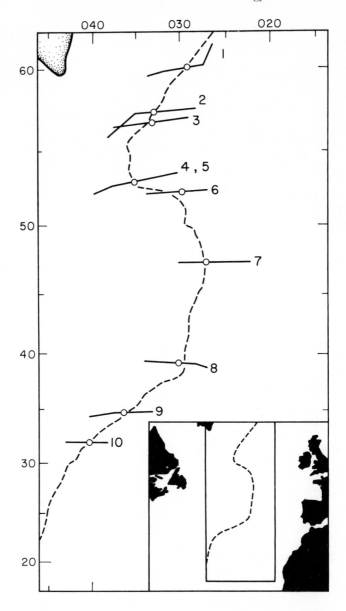

Fig. 8.6. Magnetic anomalies over the Mid-Atlantic Ridge: location of the crest of the Ridge and of flights on which the total intensity of the magnetic field was measured. The heavy broken line is the approximate position of the crest of the ridge. The circles correspond to the points on the centre line of Fig. 8.7 (after M.J. Keen[157]).

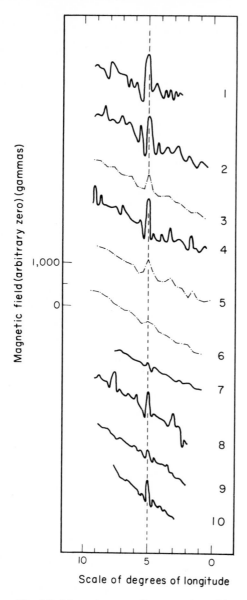

Fig. 8.7. Magnetic anomalies over the Mid-Atlantic Ridge: total intensity of the magnetic field. The profiles show the intensity of the field plotted against a scale of longitude. The zero values of the field are arbitrary; a scale in gammas is shown on the left-hand side. Numbers correspond to the flight tracks of Fig. 8.6 (after M. J. Keen [157]).

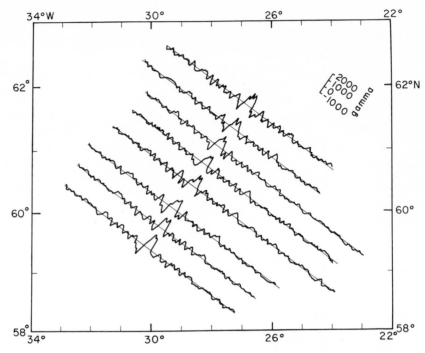

Fig. 8.8. Magnetic anomalies over the Reykjanes Ridge (Mid-Atlantic Ridge) south of Iceland. Note continuity of features parallel to the main topographic trend of the ridge (after M. Talwani and others[268]).

several hundreds of gamma when measured at sea-level, and it is considerably wider than the median valley. The rock responsible for it must therefore be wider than the median valley and not too shallow. However, rocks are not magnetic above temperatures of 500–600°C, the Curie temperatures of the magnetic minerals, and the rock responsible cannot be placed below the corresponding isotherm. Consequently it cannot be placed too deep. The coherence of the magnetic anomalies over several tens of kilometres has been established on the Mid-Atlantic Ridge south of Iceland[268] and the profiles of Fig. 8.8 show this. The dominant central magnetic anomaly is absent at fracture zones, and its spatial displacement has been used to locate such zones. Figure 8.5 shows two other features of the magnetic anomaly pattern of the Mid-Atlantic Ridge. The wavelengths which dominate the pattern for 500 km on both sides of the crest are short, and the amplitudes rather small. This zone corresponds approximately with the anomalous zone established in the seismic studies. Beyond this zone the wavelengths and amplitudes increase.[130] It seems clear that the structure of the ridge is different from that of the ocean basin on each side; perhaps it is relevant that it is at the margins of the Mid-Atlantic Ridge that evidence has been found for deformation of sediment deposited upon rough underlying topography.[81]

The cause of the large anomaly associated with the crestal region can be attributed to magnetized rock bodies a few kilometres in width and depth, with tops close to the bottom of the sea. The magnetization can be assumed to be in the direction of the present magnetic field and the susceptibility of the rock is approximately $0.005-0.01$ c.g.s. units greater than that of the surrounding rock. Such an interpretation,

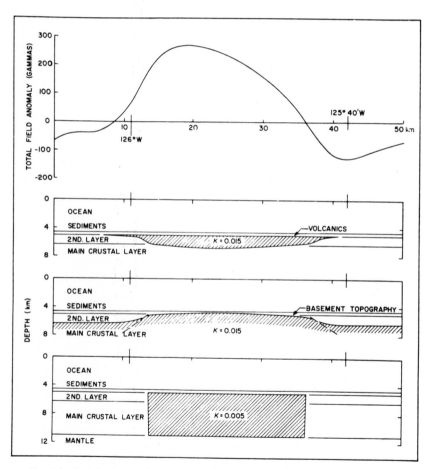

Fig. 8.9. One interpretation of magnetic anomalies found in the East Pacific. Note the magnetized body is surrounded by non-magnetic bodies, implying a difference in rock type (after R. G. Mason and A. D. Raff[190]).

similar to that shown diagrammatically in Fig. 8.9 is no doubt reasonable; the crest of the ridge is associated with vulcanism and shallow-focus earthquakes and large intrusive bodies might be expected. The profiles of Fig. 8.5 show also that anomalies of lesser magnitude persist away from the crest onto the flanks of the ridge, and these can be interpreted in a similar way. The same problem that arises in the interpretation of the north — south anomalies in the eastern part of the Pacific Ocean arises here, however (see Chapter 6), for there is no direct evidence on other geophysical grounds that the variety of magnetic and non-magnetic intrusive bodies exist, away from the crests of ridges. Before discussing the elegant solution suggested by F. J. Vine and D. H. Matthews,[284] it is worthwhile to point out that the evidence sufficient to identify such intrusive bodies must be most precise; however, that there is no correlation in detail between topography and magnetic anomalies over the mid-ocean ridges (except for the association of the large anomaly and the central valley) is evident.

Vine and Matthews have suggested that rock is transported to a region beneath the crest of the ridges from the mantle, and on cooling it acquires a remanent magnetization in the direction of the prevailing earth's field. The rock body at the crest is transported towards the flanks and new material is brought beneath the crest. If the earth's magnetic field periodically reverses, the magnetization of the rock bodies successively brought to the crest will be in opposite directions and the magnetization of adjacent bodies in the flanks will be opposed. The reality of reversals of the field, rather than self-reversals of magnetic minerals in a uniform field, has been shown in a striking way by comparing the ages of lavas collected throughout the world and their magnetic polarity. Lavas in any one age range are consistently magnetized one way, and in the succeeding age range the other way.[75] The Vine and Matthews mechanism is illustrated in Fig. 8.10. This mechanism could also account for large magnetic anomalies over abyssal plains, in places where there are no associated gravity anomalies.[175] Ruggedness in the topography buried beneath the plains would be detected by gravity measurements, and this ruggedness could account for the magnetic anomalies — so that if the gravity field is flat the origin of the magnetic anomalies must be sought in contrasts in the magnetization of the rocks which underlie the abyssal plains. One test of the Vine–Matthews hypothesis is this.[8] The "fit" of the coastlines of South America and of Africa suggest that once the two land masses were adjacent, now separated on account of drifting apart, in some way or another: to "fit" the coastlines necessitates a rotation about a point near to the Azores. Magnetic lineations run parallel to the ridge

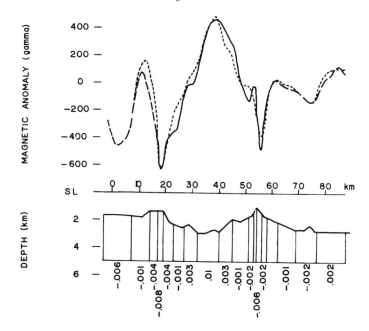

Fig. 8.10. Vine and Matthews form of interpretation of magnetic anomalies found over mid-ocean ridges. The solid line is the observed anomaly in the magnetic field. The dashed line is the calculated field corresponding to the model. This consists of blocks reversely and normally magnetized, with vertical contacts, and effective susceptibilities as indicated beneath the model (in e.m.u.). (Effective susceptibility is defined as the ratio total magnetization/earth's field strength.) Note that the adjacent blocks are not different rock types, but have different directions of magnetization (after B. D. Loncarevic and others[177]).

(Fig. 8.8) (as the solution above suggests) so that any magnetic lineation must be such that magnetic anomalies on profiles across the ridge must have a lower wave number—longer wavelength—in the south part of the South Atlantic Ocean than in the north part.

Displacements of the crests of ridges traced by off-set in the central magnetic anomaly are reflected in the pattern of the distribution of earthquake epicentres.

The shallow earthquakes associated with the crestal regions[244,262,265] are concentrated in a narrow belt only tens of kilometres wide, and in this respect differ from many continental areas of high earthquake activity; the earthquakes of East Africa are not associated only with the rift valleys, and those in Siberia beyond the Arctic Ocean Ridge are scattered, not concentrated along a line (Figs. 8.11, 8.12 and 8.13). Earthquake activity is particularly intense where fracture zones intersect the mid-ocean ridges,[123] and the evidence presented to date suggests that it is rare for the zone of earthquakes to extend beyond the limits of the displaced central part of the ridge. An example of this is seen in the South Pacific Ocean near 55°S., in the region between 115°W. and 130°W. Observations such as this play an important part in a theory of development of faults beneath the ocean basins (transform faults) described in Chapter 9. The junction of the Carlsberg Ridge with North Africa and Arabia in the Gulf of Aden and the Red Sea has been traced by bathymetry, magnetic surveys and the location of earthquake epicentres.[193,265] One sees here the narrow zone of earthquake epicentres concentrated along the Carlsberg Ridge displaced in a right-handed sense, and bent around into the Gulf of Aden (Fig. 8.11). Sykes points out that the seismicity of the Red Sea is not concentrated along the central zone of the Red Sea but that many of the epicentres are located along the margins. He points out too that the rotation of Arabia with respect to Africa can be accounted for by movement along a fault zone defined by a concentration of earthquake epicentres.

The determination of earthquake epicentres has become a powerful tool for the location of tectonically active regions. For example, detailed studies of the bathymetry of the Arctic are difficult to make, but the most likely extension of the Mid-Atlantic Ridge north of Iceland passes through Jan Mayen, between Spitsbergen and Greenland, and extends then in a straight line across the Arctic to Siberia, parallel to the Lomonosov Ridge, but not coincident with it[263] (Figs. 8.12, 8.13). A large fracture zone may run north-east−south-west in the vicinity of Jan Mayen, and the principal concentration of epicentres is between the displaced ridge crests (Fig. 8.13).

The association of earthquakes with vulcanism along mid-ocean ridges suggests that sources of heat must be located close to the central parts of the ridges, and that values of heat-flow should be high. High values associated with the East Pacific Rise are shown in Fig. 8.3.

The high values of heat-flow associated with the crests of ridges are significant for a number of reasons. First, being concentrated in narrow zones they suggest the sources of heat are not very deep−perhaps 10 km. Were the sources deeper the zones would be wider. An interpretation of this could be that the effect of recent dyke intrusion is being seen, which would be associated with vulcanism. Second, the isotherms beneath the central parts of ridges must be elevated. This means that the magnetization of rocks in the region will be confined to a shallower uppermost layer than elsewhere, because the Curie temperature will be reached at shallower depths. If the oceanic crust is serpentinized the

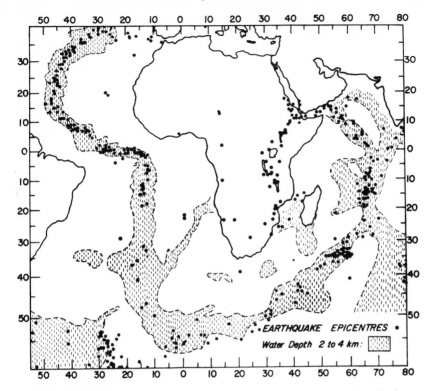

Fig. 8.11. Distribution of shallow earthquake epicentres in the South Atlantic and Indian Oceans. The seismically active belt follows closely the crests of the Mid-Atlantic Ridge, the mid-ocean ridges of the Indian Ocean, the Gulf of Aden, the southern part of the Red Sea and the area of the rift valleys of East Africa (after R. W. Girdler[97]).

base of the serpentinized layer will be higher beneath the central zones. Third, and most difficult to explain, the sources of heat have to be accounted for. One mechanism for transporting the heat is that of convection currents, postulated to rise beneath the central parts of ridges. These could also lead to lateral crustal transport away from the crests of ridges and variations in the rate of transport account for the faults associated with the ridge system.

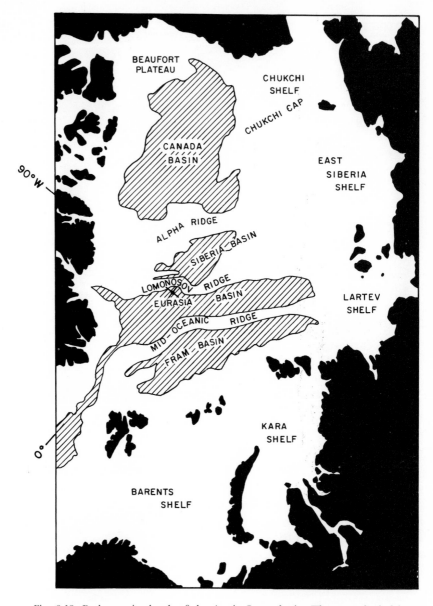

Fig. 8.12. Bathymetric sketch of the Arctic Ocean basin. The area shaded is deeper than 3000 fm, the area left blank is shallower than 3000 fm (after B. C. Heezen and M. Ewing,[121] simplified).

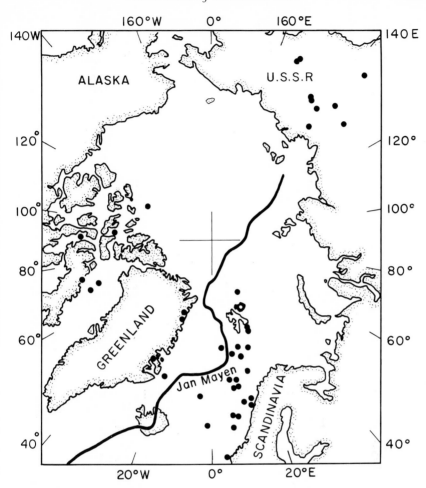

Fig. 8.13. Earthquake epicentres in the North Atlantic and Arctic Oceans. The solid line contains the majority of epicentres. Those which fall outside the line are shown as separate dots (after L. Sykes[263]).

THE OCEAN BASINS

Seismic refraction measurements in the ocean basins, away from mid-ocean ridges and continental margins show that the structure of the oceanic crust may be expressed in terms of layers (see Table 8.1).

Table 8.1

Layer	Thickness (km)	Compressional wave velocity (km/s)
Sea water	4·5	1·5
Layer 1	0·45	2
Layer 2	1·75	4–6
Layer 3	4·7	6·71
Layer 4	—	8·09

This structure is that deduced by M. N. Hill[136] in 1957; a more recent analysis by R. W. Raitt[233] is essentially the same, as Raitt points out (see Table 8.2).

Table 8.2

Layer	Thickness (km)	Compressional wave velocity (km/s)
Layer 2	1·71 ± 0·75	5·07 ± 0·63
Layer 3	4·86 ± 1·42	6·69 ± 0·26
Layer 4	—	8·13 ± 0·24

(The uppermost layer 1 has been omitted.) We shall now consider what the composition of each (seismically defined) layer may be, and whether or not the discontinuity in velocity between any two layers, tacitly

Table 8.3

	Wave velocity	
	Compressional (km/s)	Shear (km/s)
2·1	5·26	3·22
30·5	6·10	3·68
—	8·11	4·53

assumed above, is real. First, however, it is worthwhile to recall that the continental crust is very different; one crustal section beneath the Atlantic Coast of Nova Scotia was found to be as shown in Table 8.3.[11]

Some continental crustal sections would show a layer with compressional wave velocity between 6·10 and 8·11 km/s — the "intermediate" layer, bounded above and below by the "Conrad" and the Mohorovičić discontinuities respectively. A comparison of the oceanic and continental crusts shows that the depths at which the velocity near 8 km/s is attained, defining the depths to the Mohorovičić discontinuity, are very different.

Layer 1. Direct observation shows that unconsolidated sediment covers a large part of the deep-sea floor, and layer 1 can be assumed to represent these sediments. The thickness of sediments is of great imtance.[161] The rates of sedimentation in deep-sea sediments found from studies of cores some few metres in length vary from values of 0·5 mm per thousand years to some centimetres per thousand years; it is interesting to compare these rates with the actual thicknesses found by seismic profiling.[81] On the Mid-Atlantic Ridge the sediment thickness in the depth zone where carbonates may accumulate is only 100–200 m, and where "red clay" may accumulate is of the order of 50 m. The basins which border the ridges contain up to several kilometres of sediment. If the present rates of sedimentation held in the past then the thin veneer of sediment found on the ridge can respresent only a few million years of time, not long. The rate may have been smaller in the past; or sediment may have been transported off the ridges in an unknown fashion; or the reflector at the base of the sediments is only an uppermost lava flow or intrusion, and beneath it lie more sediments. However, thin sediments on the ridges can be accounted for elegantly if the oceans spread apart from the crests of ridges, in the way described in an earlier section of this chapter. The crestal parts of the ridges will then be the youngest parts.

The compressional wave velocity in layer 1 increases with depth.[79,136] Reflections from the sea bottom show that either the density, or the velocity, or both of these change across the sea-sediment interface; refracted arrivals which have travelled along the sea floor are not observed, which can mean that the velocity in the uppermost sediment, immediately beneath the sediment-water interface has a lower velocity than has the water. Studies of reflections at various angles of incidence show that at large angles, reflections which might be expected from the base of layer 1 come not from there, but from some point within the layer itself, and the apparent reflection is due to bending of the ray by the velocity gradient.[79] Laboratory studies of sediments show that

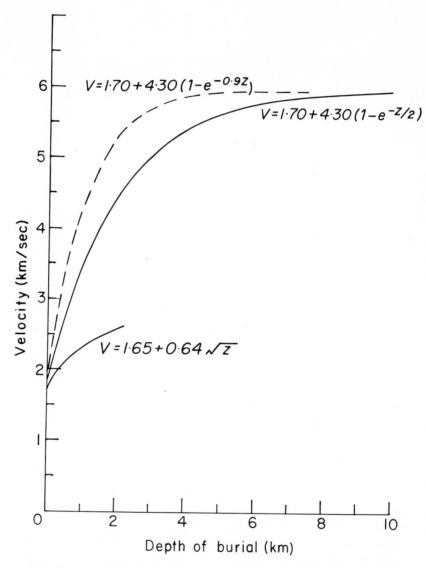

Fig. 8.14. The compressional wave velocity in sediments as a function of depth of burial, derived from seismic refraction data. The lowermost curve is fairly representative for deep-water sediments of the Atlantic, the middle curve of shallow-water sediments and the uppermost curve of the Blake Plateau (after J. E. Nafe and C. L. Drake[211]).

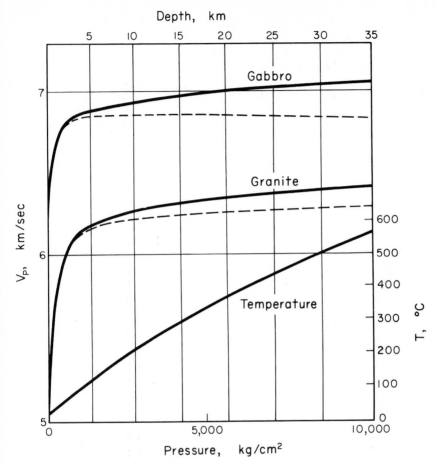

Fig. 8.15. The velocity of compressional waves for average granite and average gabbro as function of pressure and depth. The solid curve gives velocity versus pressure at 20°C. The broken curves give velocity versus depth after correction for the assumed temperatures shown by the bottom curve (after F. S. Birch[18]).

upon compaction the values of velocity increase,[166] and this is summarized in Fig. 8.14, which shows the change in compressional wave velocity with depth of burial.[211] This shows clearly the difficulty of identifying rock type from the velocity; Fig. 8.15 shows the compressional wave velocity in granite and gabbro as a function of depth.[19]

The effect of compaction of sediments in the deep-sea basins is to decrease the porosity and concomitantly to increase the velocity within the sediments.[109] The decrease in porosity followed subsequently by distortion of the grains themselves and by recrystallization will cause a calcareous ooze to change into a limestone and a clay to change into a shale, and the thickness of the originally unconsolidated sediment is reduced; Hamilton estimates that if a sequence which consists now of unconsolidated clay at the top and of shale at the bottom is at present 1 km thick then about 2·2 km of sediment has been deposited overall.

We see then that there will be a velocity gradient within layer 1, which consists of unconsolidated sediments; we might expect that layer 1 will merge into layer 2, if layer 2 consists of consolidated sediments.

Layer 2. The velocity of propagation of compressional waves in layer 2 is in the range 4–6 km/s and this large range reflects no doubt the difficulty of observing it by first-arrival information in refraction measurements, and also the number of rock types which could, and no doubt in various places do, constitute it. A number of rock types could have compressional wave velocities in the range 4–6 km/s and among them are consolidated sediments (shales and limestones), serpentinites, weathered lavas and granites. There is rather little evidence that granite can form a substantial part of the oceanic crust and we will for the moment put the thought aside, but there is ample evidence that both consolidated sediment and weathered lava-flows could form layer 2. Matthews[191] has shown that weathered lavas could form layer 2 beneath the Iberian abyssal plain; consolidated sediments outcrop on the north wall of the Puerto Rico Trench,[131] and semi-consolidated sediments have been dredged in Theta Gap.[125] The results of seismic-profiling across the Atlantic Ocean[81] show that beneath a rather thin veneer of acoustically transparent sediment occurs a highly reflective rock surface; this could be formed of many rock types, and its importance lies in the questions one asks about it — Is it the first lava-flow or intrusive igneous rock beneath the sediment? Does the sediment column above represent the complete sedimentary geological section? It seems likely that the geology of the sea floor will prove to be as complicated as land geology and that no one answer will be sufficient.

Layer 3. Layer 3 has been popularly called basaltic or gabbroic and identified as continuous with the "intermediate" layer found in many continental crustal sections, bounded above by the so-called Conrad discontinuity and below by the Mohorivičić discontinuity. This identification is based on the abundance of basaltic igneous rock in the ocean basins and the velocity of propagation of compressional waves in layer 3 and in the "intermediate" layer. Steinhart and Meyer[259] have shown that the "intermediate" layer is not as ubiquitous as was thought at one time when there was little evidence, and in the section under the Atlantic Coast of Nova Scotia of Barrett and others, described earlier,[11] it is not apparent as a discrete layer from the information provided by travel times. Direct sampling of the ocean floor has been limited, as pointed out in Chapter 7, but among those rocks described have been the dunites of St. Paul's Rocks in the Atlantic Ocean, basalts from the Iberia abyssal plain and serpentinized peridotite from the north wall of the Puerto Rico Trench.[131,191,274] All these rock types, and others, including gabbros, could have a compressional wave velocity in the range found in refraction experiments. Limestones could also have velocities within this range, and can be excluded only through apparent rarity in samples. Other rock types have also been found, metamorphic rocks among them, which will in places comprise layer 3.[195]

The compressional wave velocity of $6 \cdot 7$ km/s or higher is too great to allow granite to be postulated as a general constituent of layer 3 (see Fig. 8.15) and granite has either not been found in dredge collections or has been dismissed as "erratic" — that is, transported from elsewhere. Evidence at the present time is that acidic rocks in the ocean basins, such as granites, are either "micro-continents" — the Seychelles, for example, or occur in minor quantities as the end products of differentiation of basic magma.[222]

The Mohorovičić discontinuity and the upper mantle

The nature of the M discontinuity is of some importance.[308] The information available from crustal seismic refraction experiments that has been used in studies of the oceanic crust and of the M discontinuity under the oceans is the travel-time associated with a particular path of propagation at a known range between source of energy and detector. The plot of time against range is, it is found, a series of points which appear to fall upon straight lines. Straight lines are forced through the points and the reciprocal of the slope is the velocity of propagation in the "layer" which corresponds to any one line. The procedure was described in Chapter 2. M. N. Hill[136] has pointed out that at long ranges the ray path is through deeper parts of the layer than at short ranges, and if the velocity is increasing with depth then unless the increase is sufficient to produce a new line on the time–distance plot the points obtained at long ranges, associated with a higher velocity, will be forced through the same line as those points obtained at short ranges. The line will have a smaller slope and a larger intercept on the ordinate (the time axis) than it would if the long range points had been omitted. The consequence of this is that layer 3 will be apparently thicker, and the velocity of propagation attributed to layer 4 apparently greater. Hill pointed out too that if layer 3 is thick, then the layer 4 (upper mantle) velocity may be greater because of the increase in pressure. Some support for these comments was found in his analysis of the determinations of the thickness of the oceanic crust in the Pacific and Atlantic oceans.

One should observe too that the interpretation of time–distance curves in the form that they are usually obtained is not unique. It is true that if the surface travel times are completely known then the vertical velocity gradient can be determined, but the time–distance curves obtained have seldom been complete, and in any case, geological uncertainties intervene to produce apparent errors (apart from errors in the observation of time and distance). The most that can be said concerning M from observations of time and distance in any one seismic experiment in the ocean basins is that at some depth (often of the order of 10 km) beneath the ocean surface the compressional wave velocity reaches values close to 8 km/s. We know little, at present, from observations of travel-times, of the sharpness of the Mohorovičić discontinuity between layer 3 and the upper mantle.

To identify the rock type beneath the oceanic crust or beneath the continental crust is not easy. The major considerations are these:

(1) the rock must have compressional and shear wave velocities of the order of $8 \cdot 1$ km/s and $4 \cdot 5$ km/s respectively at the pressure and temperature which will be met just beneath M;

(2) its composition must be compatible with that of rocks actually observed on the surface of the earth;

(3) its content of radioactive elements (principally potassium, uranium and thorium) and its thermal conductivity must lead to acceptable values of heat-flow, measured at the surface of the earth;

(4) its density must be consistent with the demands of the seismic velocities and with the demands of the gravity field of the earth.

The values of heat-flow at the surface of the earth show large variations in particular regions, but are on the whole rather uniform. The heat-flow from the floors of the Atlantic Ocean and the Pacific Ocean is approximately the same as that from the continental crust.[37,237] An excellent compilation of heat-flow data is that of W. H. K. Lee,[169] who finds that the world average heat-flow is about $1 \cdot 5$ μcal/cm²s, and that the difference in average heat-flow for continents and ocean basins is small and probably not significant (with the present data). The implication of this[47,169,186] is that the mantle beneath the oceans must differ from that beneath the continents; the continental crust is radioactive (by comparison with the oceanic crust) and much of the heat-flow observed on the continents can be accounted for by the generation of heat within the continental crust itself. This is not the case in the ocean basins; even if the oceanic crust were composed wholly of rock with the same radioactivity as granite most of the heat observed at the ocean floor must have originated beneath the oceanic crust. This leads to thermal inhomogeneities between the oceanic and continental mantles, which may explain in part the distribution of active volcanoes and of earthquakes.[186] The thermal inhomogeneities will lead also to differences in density, implying that equality of mass per unit area over the earth demands vertical sections taken to the depths of the base of the upper mantle (say 400 km), not merely to the greatest depth of the continental M discontinuity. Inhomogeneities in seismic wave propagation within the upper mantle are well established (though not systematically described[218]), and a number of studies have shown that a zone of low shear-wave velocity must exist in the upper mantle.[35,60,266]

The geological demand upon the choice of rock type in the upper mantle is principally this—that it can give rise to the large quantity of basaltic rocks which are found on the surface of the earth. There are two ways to proceed—we can choose a rock chemically different, and in large part mineralogically different from basalt, which could generate basaltic rocks, or we can choose a rock chemically identical with basalt, but with a different mineral assemblage. The chemical change could be to a rock of ultrabasic composition, or to a mixture of basalt and an ultrabasic rock (such as dunite). The basalt–dunite mixture was called "pyrolite" by S. P. Clark and A. E. Ringwood.[47]

Suppose (to illustrate the difficulties) that peridotite and eclogite are chosen as alternative constituents of the upper mantle. Peridotite could satisfy the compositional requirement, because basalt could be generated from it, and so might be a suitable rock to consider beneath the ocean basins at least. However, it is not sufficiently radioactive[276] to account for the 55–60 ergs/cm² sec of heat which must flow into the base of the oceanic crust. It may be sufficiently radioactive to account for the 15 ergs/cm² sec of heat which must flow into the base of the continental crust. Peridotite has an additional virtue, that being olivine rich, the oceanic crust might be formed by its serpentinization, discussed in the last section of this chapter. Eclogite could also satisfy the compositional requirement, and is sufficiently radioactive that enough heat is generated beneath the oceanic crust; there are difficulties of transporting the heat— if it is transported by conduction alone the source must be confined to the upper few hundred kilometres of the earth, because of the time involved. Unlike peridotite it is too radioactive to be the major component of the upper mantle beneath continents. Eclogite is less dense than peridotite, and if proposed as the oceanic upper mantle, with peridotite the continental upper mantle, could lead to the observed lateral inhomogeneities in seismic velocities, deduced from surface wave and body wave observations, and from measurements of g[156].

The stability fields of eclogites and basalts now become relevant (Fig. 7.1). Yoder and Tilley[309] have shown that at the pressures and temperatures of the M discontinuity beneath oceans, basalt (not eclogite) is the stable phase. Eclogite might be the stable phase at M beneath continents. Consequently the scheme of the preceding paragraph is unsatisfactory, unless eclogite is not encountered for some tens of kilo- metres below the oceanic Mohorovičić discontinuity, in the region where it is stable.

Whatever the solution adopted, the lateral inhomogeneities which result from the uniform free-air gravity anomalies and the uniform heat-flow over continents and oceans remain. These present difficulties if convection currents in the mantle are proposed. Convection currents are attractive as a heat transport mechanism in the upper mantle where radiative transfer of heat cannot take place because the temperature is too low, and as agents in continental drift.

The oceanic crust and serpentinization

The two types of rock which are the most likely to form the main part of the oceanic crust are basic rocks (gabbros and basalts) on the one hand, and serpentinites (serpentinized ultrabasic rocks) on the other. At the present time it is certainly not clear which (if either) predominates. The two are fundamentally different. Basalts and gabbros can be generated by fractional crystallization or partial melting from an ultrabasic source, by phase conversion from eclogite, or by melting of a basaltic rock. The last is unlikely, because it implies that the upper mantle is basaltic, a suggestion which can be rejected for a number of reasons, of which the velocity–pressure relationships are one. Either the first or second mechanism for generation of basalt is reasonable, and leads to the picture of the oceanic crust being composed in large part of this rock—a "layer 3" several kilometres thick. Serpentinites arise through hydration of ultrabasic rocks, not of eclogites (of basaltic composition), nor of course basalts or gabbros, but rocks which consist primarily of olivine, and which can be the source of the basalts or of the eclogites under suitable conditions of pressure and temperature. If eclogites are proposed as upper mantle rock then basalts must be generated by phase changes, and ultrabasic rocks must be the residue of differentiation of the basalts themselves.

Serpentinite is found in dredge hauls from fracture zones of the Mid-Atlantic Ridge, from the north wall of the Puerto Rico trench, from St. Paul's Rock in the Atlantic Ocean, and is well known in ultrabasic intrusions on the continents, and on islands of the Caribbean island-arc system and elsewhere.[134] The association of the serpentinites on land may be with fold mountains, for example the Appalachians, or in the cores of folds in Puerto Rico, where they are associated with volcanic rocks and with marine sediments. They are also well known as massive intrusions—of which the Muskox intrusion of the north-west Territories of Canada is a dramatic example.[85] One question which is pertinent is to ask whether the serpentinization has taken place through the action of sea water from above, or through juvenile water in the mantle itself. If the former, then the case for layer 3 being largely serpentinized upper mantle is made weaker; naturally both could have occurred from place to place, and one answer will not necessarily be the whole truth. The Muskox intrusion outcrops in Precambrian metamorphosed sediments and igneous rocks and is 1155 million years old. It is funnel-shaped in cross-section, longer than at least 90 km, with a long feeder-dyke (the stem of the funnel); the depth of the funnel is about 2600 m, and its diameter about 10 km. The vertical length of the feeder now exposed is about 5 km. Three holes have been drilled into it which effectively sample the whole of the funnel at its centre and at one side. The rocks which make up the major part of the funnel—the central layered series—are, generally speaking, ultrabasic towards the bottom—dunites and serpentinite among the rock types, gabbros towards the top which become richer in acidic rocks (such as granophyres) upwards. The serpentinization of the dunites is extreme through most of the intrusion, except towards the bottom of the funnel, implying that serpentinization was from the top downwards. The point here is that in a core effectively 2600 m long it is only towards the bottom that serpentinization becomes less intense. A hole drilled into a serpentinite body in Puerto Rico[44] was only 305 m long, which if the Muskox intrusion is a guide is not long enough to penetrate below a zone of serpentinization caused by water from above. Perhaps the Muskox intrusion is not a fair comparison; however, a similar feature has been noticed in another ultrabasic intrusion—the Blue River intrusion in British Columbia,[305] and serpentinization is found to be most intense at its margins.

A serpentinite layer 3 will be less magnetic[44] than the normal basalt. This means that in any calculations which attempt to simulate the measured magnetic field over the oceans, the crustal structure which must be assumed is of the following type: water—5 km; sediments—0·5 km; basalt (magnetic)—1·5 km; serpentinite (non-magnetic)—4 km.

In one set of calculations Vine found that such a structure was, in fact, preferable to one in which the whole oceanic crust, except sediment, consisted of (magnetic) basalt.[304]

The question of the serpentinization of layer 3 seems unresolved.

THE CONTINENTAL MARGINS

Their Significance

The boundary between the continents and the oceans separates regions different in every way one from another. The continents and ocean basins have different geological features, the Mohorovičić discontinuity changes elevation from about −35 km to about −10 km and the continental crust apparently vanishes.

The borders of the continents are the site of a large part of the world's vulcanism and of many earthquakes. Thermal disequilibrium must exist between the upper mantle beneath the oceans and the upper mantle beneath the continents.[186] C. L. Drake and others[62] in a stimulating account of the continental margin of the east coast of North America have emphasized the requirement for heat at the margins, if the present margin is to be converted into the equivalent of an ancient geosyncline.

The Structure at the Margins of Continents

The margin which has been studied in greatest detail is that of the east coast of North America. Whilst the detail available is of great value, it must not be thought that any conclusions reached which concern this margin can necessarily be supposed to apply to all margins.

The shelf is underlain by a thick section of sediments, bounded near or beyond the shelf break by a rise in the basement which underlies the sediment.[156] The thickness of this section is about 5 km under Sable Island, for example, off Nova Scotia.[16] On the seaward side of the basement rise the sediments thicken again, so that under the continental rise they may be several kilometres thick, beyond which they thin towards the ocean basin. It is not easy to identify the nature of the basement beneath the sediment, either beneath the shelf or beneath the continental slope and rise. Drake and others have pointed out that the compressional wave velocity in the basement decreases away from the continent and have related this to decrease in metamorphism away from the centre of the Appalachian system. Dalhousie University and

Lamont Geological Observatory have studied the relationship between the continental margin and the Appalachian System. This Palaeozoic mountain range extends the length of eastern North America. Any extension would cut the continental margin northeast of Newfoundland at right angles, and in this the System bears the same relationship to the margin as the Hercynian mountain range of the late Palaeozoic does to the continental margin off western Europe. The question asked in both cases is "does the ancient mountain range extend beyond the margin of the continent?" It would be difficult to reconcile evidence that these continental structures which trend at an angle to the margin do extend into the ocean basin with the evidence that tells us that the crust of the ocean basin is completely different from the crust of the continents.

Magnetic surveys of the shelf and slope northeast of Newfoundland let us see what happens to magnetic anomalies which, in Newfoundland itself, trend parallel to the Appalachian System. Large magnetic anomalies are indeed found over the shelf and slope northeast of Newfoundland, but they trend parallel to the margin, not parallel to the Appalachian System. This suggests that the Appalachian System does not cross the boundary between the continental crust and ocean basin. A similar study was made of the Western Approaches to the English Channel.[139] This survey shows that the easterly trending magnetic anomalies associated with Hercynian structures of western Europe die out before the continental margin is crossed.

The magnetic anomalies which run parallel to the edge of the continental shelf of eastern North America may be characteristic of many continental margins. They could arise as an "edge effect". If the continental crust to the depth of the Curie point isotherm is more magnetic than the oceanic crust and mantle to corresponding depths, we would expect broad anomalies over the boundary between the two regions.

The sediments at the foot of the continental margins are derived from the continents and transported by turbidity currents. Cores taken from the continental rise, and from the abyssal plains beyond, show characteristics like those which have been described from the Iberian abyssal plain (Chapter 5); alternating coarse-graded beds and pelagic sediments, derived shallow-water fauna in the graded beds, and a sparse macro-fauna. These are like those in the part of the Appalachian geosyncline called the eugeosyncline, in which thick sequences of alternating graded beds and shales are common, and which are associated with vulcanism. The sediments beneath the shelf of the coastal plain are of a shallow-water facies, and these may resemble the miogeosyncline of the Appalachian system. The basement ridge which separates the two sedimentary sequences corresponds to the Precambrian

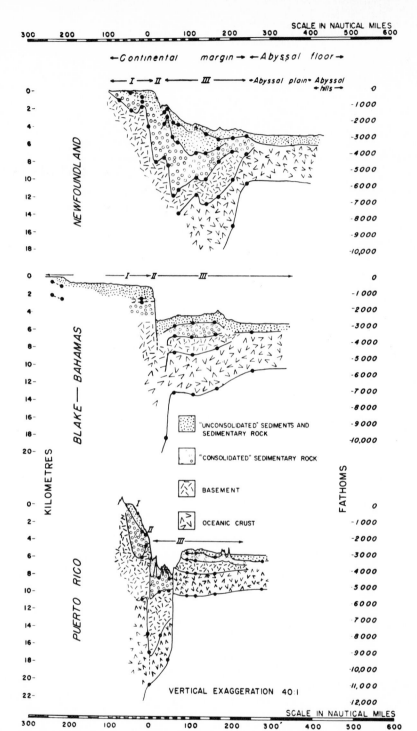

Fig. 8.16. Crustal structure of the continental margin of eastern North America. The dots represent depths determined in seismic experiments. The different rock types are based upon interpretation of the seismic velocities (after B. C. Heezen and H. W. Menard[126]).

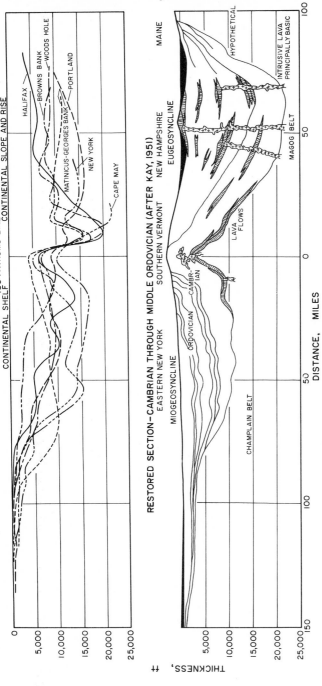

Fig. 8.17. Comparison between the Appalachian geosyncline, restored as it may have been in Cambrian to Middle Ordovician (after M. Kay), and lines of equal sediment thickness off the east coast of North America (after C. Drake, M. Ewing and G. H. Sutton[62]).

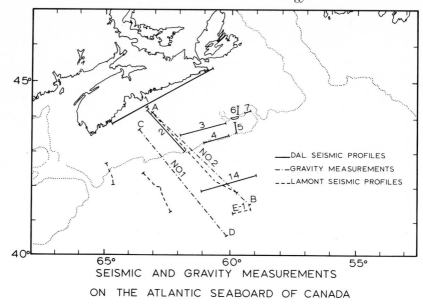

Fig. 8.18. Continental margin off eastern Canada: location of seismic profiles and ship's tracks along which gravity observations have been made (after C. E. Keen and B. D. Loncarevic[156]).

ridge which separates the Appalachian miogeosyncline on the east. These features are seen in Figs. 8.16, 8.17, 8.18 and 8.19, which show profiles off the eastern seaboard of North America, and a section across the Appalachian geosyncline.

The resemblances between the continental margin of eastern North America as it is now and the Appalachian geosyncline are many, and the importance of the ideas put forward by Drake and others cannot be underestimated; they lead to a number of other comments. The presence of a thick sequence of graded beds (with intercalated pelagic shales) implies that there was a considerable supply of sediment to hand. The graded greywackes of the Appalachian system, and of the Lower Palaeozoic geosyncline of Great Britain must have been at a considerable depth beneath the sea such as the sediments of the continental rise, and the sediments of the basins off southern California[86] are today. Further, we know that in places beneath the graded greywackes of the continental geosynclines are found basal conglomerates resting upon a metamorphosed basement; this is seen, for example, near Salrock in County Galway, Ireland, where Silurian rocks overlie Connemara schists. The Connemara schists led a hectic life, even after metamorphism—they were eroded (the basal conglomerate), and subsequently

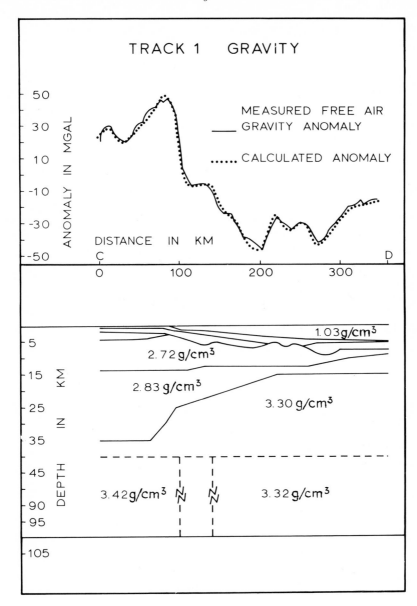

Fig. 8.19. Continental margin off eastern Canada: interpretation of seismic and gravity data. The model shows the densities assumed to satisfy the observed gravity. Note the change in density in the upper mantle (after C. E. Keen and B. D. Loncarevic[156]).

rested in deep water with some thousands of metres of greywackes above them. This being so for ancient geosynclines, it must mean that parts of the continental crust (represented by the Connemara schists) extended some way from the edge of the continental shelf. If we examine other continental margins, other curiosities come to light. For example, the flysch of the region around Nice in southern France needs a source where there is now deep sea,[162] and what is now deep sea must once have been a part of the continental land mass. The suggestion that beneath the continental rise may lie "basement", in the sense that the Connemara schist referred to above is basement, leads to the conclusion that although the schist and the overlying greywackes may at one time have been at some considerable depth below sea-level and the schists were overlain by a few kilometres of sediment, in the process of elevation into a mountain range the degree of metamorphism which they suffered was rather slight.

The transition from continental to oceanic crust is illustrated in Figs. 8.18, 8.19 and 8.20. Figure 8.20 shows seismic sections through various parts of the Appalachian System;[76] although at first sight it seems that the seismic studies reflect the two-sided symmetrical nature of the Appalachian geosyncline (a concept developed recently by H. Wil-

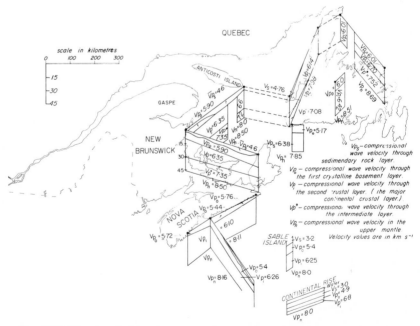

Fig. 8.20. The Appalachian System: sections to show the structure and compressional wave velocities (after G. N. Ewing and others[76]).

liams[295]), the situation may be more complicated because of thick intrusive granite batholiths. Figure 8.19 shows the continental margin itself. The most far-reaching conclusion suggested by this diagram is that the deep structure beneath the crust of the continents differs from that beneath the ocean basin, by being more dense. This is in agreement with conclusions reached by considerations of heat-flow data.[186] A different type of transition is found on the west coast of Canada. The continental shelf is narrow and the area tectonically more active than the east coast. The mountain ranges inland are relatively young by comparison with the east coast's Appalachian System. Off the continental margin may lie a short section of the East Pacific Rise, displaced by a "transform fault" along the west coast of the United States.[304] Consequently, the crust and mantle beneath this margin might be different. The study by White and Savage[292] in British Columbia shows that it is. Instead of the "normal" or higher than normal compressional wave velocities found in the mantle beneath the Appalachian System of eastern Canada, the velocities in the mantle beneath British Columbia are lower than normal; this, Lambert and Caner[165] suggest, is in agreement with estimates of electrical conductivity in the mantle, and they attribute the low compression wave velocities to higher temperatures in the mantle.

These comparisons illustrate the fact that continental margins are not all the same: some are relatively stable — like that of eastern Canada, and some are fractured into basins and troughs and associated with tectonically active regions, defined so by the location of earthquakes — that of southern California. The margins of the Pacific Ocean are very different in general from those of the Atlantic Ocean, being associated with island arcs and trenches, and transcurrent faults parallel to the margins. The structure of the crust beneath each margin is different, too. The Pacific coast of North America has a crust which thins rather gradually from continent to ocean basin,[252] whereas beneath the Atlantic coast of North America the change in crust is abrupt.

Sediments on the Continental Margin off Florida

The continental margin about which most is now known from the point of view of stratigraphy is that off Florida, where six holes have been drilled to depths below sea bottom of as much as 320 m in water depths as great as 1032 m.[43] The continental shelf off Florida breaks at about 100 m into the Florida–Hatteras slope, and at 800 m this slope merges into the Blake plateau. The Blake plateau ends at the Blake escarpment, which acts as the continental slope proper, the boundary between the continent and ocean basin (Fig. 8.21). The plateau peters out to the north, and no comparable feature is met until the Grand Banks are reached south of Newfoundland—comparable in great width, that is. The sediments of the Atlantic coastal plain—Cretaceous and Tertiary among them—are found on land adjacent to the shelf and plateau. The stratigraphy shows that the continental slope has pro- graded—that is, built outwards—approximately 15 km since the Eocene, and that the accumulation of sediment in that time has been in the range

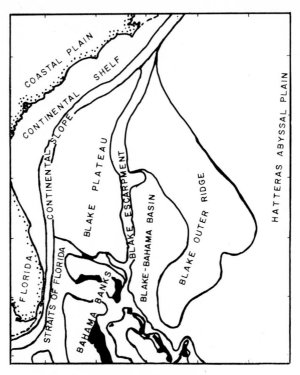

Fig. 8.21. The major physiographic provinces in the region of the Blake Plateau. The continental shelf depths are less than 300 m, the Blake Plateau level is approximately 800 m and the level of the Hatteras abyssal plain is approximately 5400 m (after R. M. Pratt and B. C. Heezen [228]).

100–200 m on the shelf and slope. The sediments were deposited at approximately the depths at which they are now found—the continental margin having subsided only slowly. The sediments on the Blake Plateau of Eocene to Recent age are thinner than those beneath the shelf, because there are unconformities of post-Oligocene and middle-to-late Eocene age, caused probably by absence of sedimentation. Bunce and others point out that the area of missing post-Oligocene strata underlies the present Gulf Stream. The slope between shelf and Blake Plateau does not appear to be a fault but arises through differences in depositional and erosional histories in the two areas. This continental margin thus appears to be a wedge-shaped constructional feature, which thins seaward. It would be dangerous to extrapolate and suppose that all continental margins are of the same nature; however, the margin off central California appears to be very similar, but in an earlier stage of development.[319]

The influence of the Gulf Stream in preventing deposition upon the Blake Plateau, and presumably also in preventing continental detritus crossing the Blake Plateau to the Outer Ridge, for example (Fig. 8.21), is very great. Heezen and others[328] have demonstrated that the major ocean currents are much more important as geological agents than has perhaps been suspected. The Gulf Stream runs north-eastward off the continental shelf of the eastern part of the United States and beneath it runs a counter current, flowing in a southerly direction. Bottom photographs of current structures show that they are aligned parallel to the appropriate prevailing bottom current. Sediment which forms the Outer Ridge to the east of the Blake–Bahama basin cannot originate off Florida, because of the Gulf Stream. It can, however, originate from more northerly areas, being transported by the deep Western Boundary current.

Island Arcs and Deep-sea Trenches

A topographic map of the ocean floors shows that the deepest parts of the oceans occur as linear or arcuate features called trenches, where the water depths may be in the range of 8–10 km, rather less than twice the depth of the major parts of the ocean basins (Figs. 8.22 and 8.23). These trenches are bordered on the landward side by mountain ranges — as off South America, where the range is the Andes, or by arcuate island groups. An example of the latter is the island arc of the Antilles, where Puerto Rico lies on the continent side of the Puerto Rico trench (Figs. 8.22, 8.23 and 8.24). The island arcs are volcanic, with single or double rows of volcanoes occurring about 250 km from the axis of the trench. Vening Meinesz was the first to find that negative isostatic gravity anomalies, centred rather on the landward side of the axis of the trench, are characteristic features of the trenches; smaller positive anomalies occur over the islands or on the continent, and on the outer (seaward) ridge of the trench.[86] "Isostatic equilibrium" means that there is equal mass per unit area beneath the earth's surface, and its chief expression is in the change in thickness of the crust from continents to ocean basins. A negative isostatic anomaly means that there is a mass deficiency

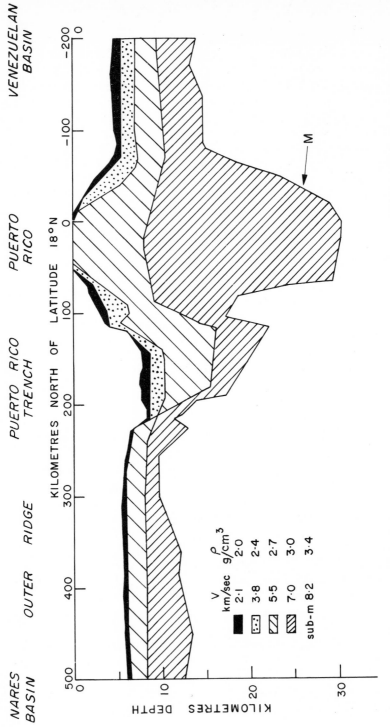

Fig. 8.22. The Puerto Rico trench: crustal structure deduced from seismic and gravity observations. The lowermost boundary is the Mohorovičić discontinuity (after M. Talwani and others[270]).

beneath; in the case of the trenches we would expect, if there was isostatic equilibrium, that the dense rocks of the mantle would bulge up to compensate for the bulge down of (light) water. There is not

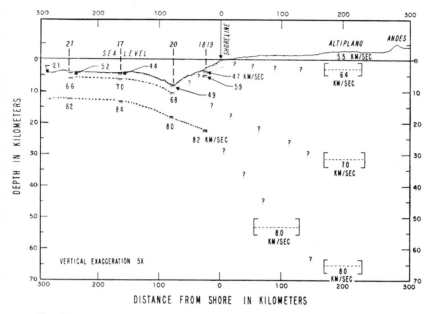

Fig. 8.23. Seismic sections across the Chile trench and adjacent Andes. The numbers show compressional wave velocities in km/s (after R. L. Fisher and R. W. Raitt[87]).

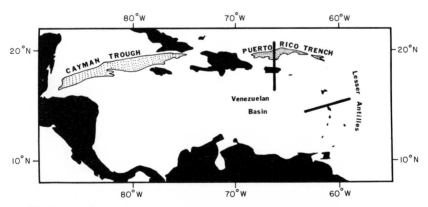

Fig. 8.24. Index map of the Caribbean. Earthquake epicentres are distributed through the arc of the Cayman Trough, Puerto Rico trench and the Lesser Antilles. Water depths in Puerto Rico trench greater than 4000 fm, and in Cayman Trough between 2000 and 4000 fm are stippled.

isostatic equilibrium, but negative isostatic gravity anomalies, implying a mass deficiency, and so we anticipate that the crust beneath the trenches is thicker, not thinner, than that of the surrounding ocean floor. Seismic refraction experiments have shown this is so.[270]

Earthquakes are associated with the trenches and island arcs,[105] and there is evidence that the depth of focus of these increases from the trench axis towards the island arc or mountain chain. Shallow earthquakes, with a focal depth less than 70 km occur rather shoreward of the axis of the trench, earthquakes with focal depths in the range 70–300 km beneath the island arc or mountain range, and deep earthquakes of focal depth 300–700 km under the basin inside the island arc, or beyond the mountain range, on the side away from the trench. However, studies made by Sykes[264] of the seismicity of the Caribbean cast doubt upon this apparent simplicity. A zone of earthquakes runs from the Gulf of Honduras off central America through the Greater Antilles (Hispaniola and Puerto Rico) and Lesser Antilles to Venezuela (Figs. 8.24, 8.25 and 8.26). The distribution of the earthquakes with regard to depth and size is not particularly simple. Although across the southern part of the Lesser Antilles the earthquakes fall along a surface which dips at 60° towards the Caribbean, in the northern part the distribution is more complicated. The earthquakes associated with the Puerto Rico trench are not concentrated along its northern wall, or the deepest part, but rather beneath its southern wall—that furthest from the Atlantic Ocean basin proper. Sykes and Ewing point out that this observation may mean that the southern wall is the more active of the two; this is in accord with the observations of Ross and Shor[243] of the Middle America trench. They suggest that the active fault there is that at the foot of the continental slope, the trench itself being a block of sea floor tilted downwards and inwards towards the continent.

Seismic reflection experiments completed recently over the Japan trench by W. J. Ludwig and others show that the seaward wall of the trench—corresponding to the north wall of the Puerto Rico trench, and the western walls of those off South America—is dominated by a host of steeply dipping faults. By contrast, the normal oceanic crust on the seaward side is not cut up by faulting. The compressional wave velocities in the crust beneath the Puerto Rico trench, the ocean basin outside the trench and under Puerto Rico itself are rather high by comparison with more "normal" areas[86,270] (see Fig. 8.22), and the velocity in the mantle beneath the trenches and the adjacent areas is about 8·0 km/s. The seismic refraction experiments suggest that the Puerto Rico trench is bounded by faults approximately coincident with its walls which run along the length of the trench, and the section of Fig. 8.22 makes it

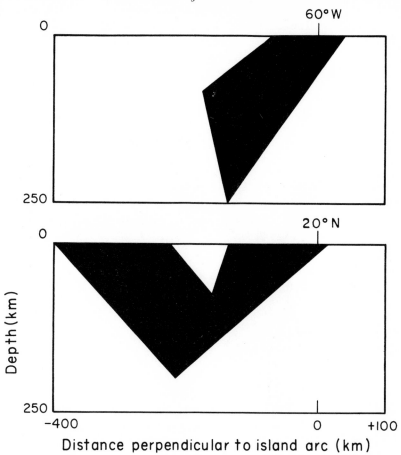

Fig. 8.25. Distribution of earthquakes beneath Greater Antilles (lower) and Lesser Antilles (upper). Lower: N.–S. section through Greater Antilles, the black enclosing all well-determined earthquakes between 64°W. and 69°W. Upper: E.–W. section through Lesser Antilles, the black enclosing all well-determined earthquakes between 14°N. and 16°N. (after L. Sykes and M. Ewing[204]). The centre line of each area is shown in Fig. 8.24.

appear as if the oceanic crust of an otherwise normal continental margin had dropped down because the crust was pulled apart — a tension hypothesis, or pushed or pulled down — a compression hypothesis. Values of heat-flow in the Peru–Chile trench are rather lower in the trench than elsewhere,[132] and this has been taken to support the hypothesis that a convection current which rises at the crest of the East Pacific Rise is sinking beneath the Peru–Chile trench. This suggests that sediments at the bases of the trenches should be deformed, by the same motion, but

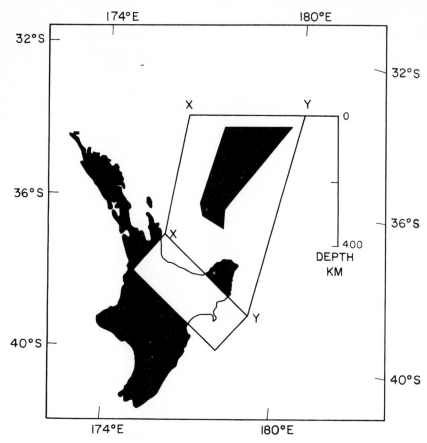

Fig. 8.26. Earthquakes beneath the Northland, New Zealand, sub-crustal rift. All earthquakes beneath the area of the rectangle have been projected onto a single cross-section. The black encloses all the earthquakes. X,X and Y,Y correspond. Earthquakes of uppermost 40 km omitted (modified from G. A. Eiby[311]).

those in the Middle America trench at least are not.[243] The thickness of sediment in some trenches is less than one might naïvely expect — the walls of the Philippine trench if projected downwards meet only 300 m below the present bottom. Hess has postulated that serpentinized peridotites should be associated with the trenches, as the crustal rocks, and some have been found in the Puerto Rico trench.[131]

There are two groups of theories which concern the origin of trenches. H. H. Hess[133] suggests that the isostatic negative gravity anomalies imply that the centre block would bounce up again in a short period of time, if it had only fallen down in a crust pulled apart under tension. It must be kept down by forces acting inwards upon the block. Compression can be produced by convection currents in the upper mantle, of various configurations, which turn downwards at the trenches. Hess proposes that a serpentinized region from the ocean basins brought down beneath the trenches will be de-serpentinized (dehydrated) at temperatures near 500°C, in accord with the experimental evidence described in Chapter 7. Similarly, volcanic rocks and sediment brought down from the continent or island side will melt. The magmas which result escape as volcanoes on the continent or island side, rather than the ocean basin side, because vertical fractures perpendicular to the arc develop preferentially on the concave side of the system; this preferential fracturing arises because of the extension of the arcuate crust horizontally as it moves towards the trench from the one side, and the compression as it moves towards the trench from the other side. This or other similar mechanisms would certainly lead to mixed magmas and perhaps to the andesites sometimes found in island arc-trench systems or continent-trench systems. A second and simpler mechanism for the origin of trenches arises from the observations of near-surface faulting on the seaward walls, and the thickening of rock of relatively low compressional wave velocity on the landward side of the trench. This can be seen in Fig. 8.22 where the rock of velocity 5·5 km/s dominates the crust beneath the southern part of the Puerto Rico trench. Ludwig and others suggest that in the case of the Japan trench the seismic and bathymetric observations can be explained if the landward wall is being built outwards — extending the axis of the trench seaward and lessening its maximum depth. The load of the island arc depresses the oceanic crust — represented by the trench proper and its seaward wall, and leads to faulting in the bent plate of crust — observed in the seismic reflection experiments on the seaward wall.

Studies of earthquakes associated with trenches are valuable for two reasons: they define the unstable zones and also allow an estimate to be made of the movement associated with the faulting which generates the earthquakes.[14,140]

CHAPTER 9

Polar Wandering and Continental Drift

The hypothesis that "polar wandering" has occurred supposes that there has been relative displacement between the present configuration of the surface of the earth and the earth's axis of rotation. The hypothesis that "continental drift" has occurred supposes that there has been relative displacement between the continents. Continental drift implies polar wandering, polar wandering does not imply continental drift.

There is no evidence which of itself proves or disproves either hypothesis. Much data can be selected which points towards one or the other as explanation. The evidence is of several sorts:[141] the shape of the continents,[24,39] the geology of complementary features,[189] the inferences which can be drawn from geology which concern climates in the past,[215] and inferences which can be drawn about past geography from a study of the magnetization of rocks.[51,245,247]

One fit of the continents which border the Atlantic Ocean is shown in Fig. 9.1. This is a fit which was found quantitatively by Bullard, Everett and Smith.[39] Similar fits can be made for the continents of Australia, Africa and India. In the particular fit shown in Fig. 9.1 geological licence has been taken to omit Iceland, and to rotate Spain towards France. Iceland has been omitted because of evidence which suggests that it began to develop only at the time the continents began to move apart. The rotation of Spain is justified by palaeomagnetic evidence, which suggests that it occurred after the Triassic period.[148] This sort of fit leads naturally to ask if the continental boundaries fit so well, does the geology not fit well too?[189] It is easy here to be biased and choose only what one wants to see, but there are certainly similarities between one continent and an adjoining continent. For example, the Appalachian mountain system runs the whole length of the eastern seaboard of the United States, into the Maritime Provinces of Canada and Newfoundland. This system is a mountain belt developed in a geosyncline of Palaeozoic times. Running through the north-western parts of the British Isles are the remnants of a mountain system developed from a geosyncline of the same age. Both systems, if continued to their continental margins, would intersect them perpendicularly, approximately. There is geological and geophysical evidence showing that neither peters out towards the margin, and no evidence has been found for their continuation across the floor of the Atlantic Ocean. They would continue one into the other quite naturally if the continents are fitted in the way shown in Fig. 9.1.

There are many similarities in the details of the two systems, the Appalachians and the Caledonides, and their matching counterparts in Morocco, Greenland and Scandinavia.[96] Across north-east Newfoundland H. Williams has demonstrated that the Appalachian system is a two-sided, symmetrical system; the margins to south-east and north-west show relatively underformed lower Palaeozoic rocks deposited unconformably upon a Precambrian basement—the north-west margin being a part of the Canadian Shield, a stable block (at this time). (There may, of course, be a part of the Appalachian system to the east of Newfoundland beneath the Grand Banks, comparable to the slates and quartzites of the Meguma Series found in Nova Scotia.) Similarly, the Caledonian system of the British Isles is two-sided and symmetrical, the forelands being on the one hand to the south-east beneath central England, and on the other hand along the north-west Highlands of Scotland. However, nowhere else are both sides of the system seen in the one region. In Morocco the reverse is seen; the margin is to the interior of the continent, as along the eastern seaboard of the United States, and there is no north-west margin. The systems of the eastern seaboard of North America and Morocco become a single, two-sided system if the continents are fitted together in the manner of Fig. 9.1. The relationships between the lower Palaeozoic rocks of Greenland and Scandinavia are similar.

There are also faunal similarities. One facies of the Cambrian rocks are distinguished by the presence of Trilobites. These can be divided into two dissimilar faunas, little intermixed, occurring on the south-east and north-west margins of both Appalachian and Caledonian geosynclines (and owing their separation to a physical or ecological barrier).

There are also differences, not yet explained if the former fit of continents is accepted. The age of the Precambrian rocks of west Newfoundland and Newfoundland Labrador are in the range of 800–1000 million years.[260] Those of the Lewisian of north-west Scotland are rather older. There are large granite batholiths in south-west England of Carboniferous age, but none in Nova Scotia or Newfoundland, so far as is known now (they may be beneath the sea, north-east of Newfoundland).

The Great Glen Fault runs in a north-easterly direction across Scotland, and it is a transcurrent or wrench fault along which there has been a lateral displacement of more than one hundred kilometres.[159] J. Tuzo Wilson has suggested that this fault has its equivalent in Newfoundland, Nova Scotia and New England, which he has called the Cabot Fault.[297] This fault must be a more complex system of faults than that of which the Great Glen Fault in Scotland is part, and no one fault dominates, as it does in Scotland.

It was at one time thought to be quite clear what climatic conditions must have been on the earth in the past. Coal measures reflected a hot and wet environment, salt deposits (evaporites) a warm dry climate. Certain assemblages of rock ressembled glacial tills and the corresponding climate was cold, sand dunes represented deserts akin to the Sahara of today. Coral reefs implied warm, tropical shallow water. However, modern studies cast doubt upon the correctness of these interpretations.[212] All that can be said with certainty of coal measures is that the vegetation demanded water to grow in, ancient "glacial tills" resemble turbidites, and coral reefs which could be mistaken for the typical tropical shallow-water coral reefs if found fossil are found in relatively deep water of high latitude.[271] What remains certain of the climatological conclusions provided by geology may be derived from the distribution of the deposits in the past; thus the distribution of evaporites seems to have been different in the past, and to have changed systematically.[179] Equally convincingly, however, Stehli has demonstrated that some faunal evidence contradicts the interpretation of the distribution of evaporites in the Permian.[257]

The fitting of the shapes of continents, the comparison of geologies and the estimation of ancient climatic conditions from geology are more or less qualitative procedures. Quantitative estimates of ancient latitudes, and of ancient geomagnetic pole positions can be obtained from the measurement of the direction of magnetization of rocks.[148,245,247]

Minerals such as magnetite and haematite acquire a strong magnetization if they are cooled from above their Curie temperature in a magnetic field. The direction of the magnetization acquired in this way is likely to be parallel to the direction of the ambient field. Thus lavas which contain such minerals can acquire a magnetization parallel to the field of the earth at the time the lavas cool. Similarly, magnetic minerals deposited chemically acquire a magnetization parallel to the earth's magnetic field.

The magnetic field of the earth at the present time is similar to that of a magnetic dipole the centre of which is at the earth's centre; the axis of the dipole is not quite coincident with the geographic axis. If the ambient field in which a rock acquires its magnetization was a dipole field the inclination of the magnetization from the horizontal is related by a simple geometric relationship to the geographic latitude. The declination or azimuth of the direction of magnetization together with the geomagnetic latitude allows a pole position to be calculated: this is the position on the surface of the earth now of the intersection of the axis of the (dipolar) magnetic field which caused the magnetization. If the assumption is made that the geographic and geomagnetic axes coincide *on average*, an assumption which can be tested for the Tertiary, then the results give the ancient geographic latitude of the rock sample, and the ancient position of the geographic poles.

It is found that rocks from the same region of the same age give latitudes and pole positions consistent with one another, but different from those of the present day; in particular rocks which were assigned on the basis of geological evidence to warm climates yield low latitudes. The position of the poles calculated from the data of rocks of the one general region, but of different ages, change systematically with time. Moreover, the results from different regions, themselves self-consistent,

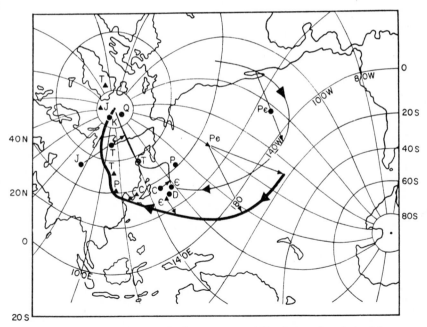

Fig. 9.2. Polar wandering paths for Europe and North America, derived from the mean poles. Triangles, American rocks; circles, European rocks. The letters indicate geologic time, thus T for Triassic (after S. K. Runcorn[247]). Reproduced by permission of the Royal Society.

may differ systematically. Thus the curves drawn on the surface of the globe of the poles calculated from the data of European rocks and from the data of North American rocks differ, the North American curve lying to the west of the European curve. This is shown in Fig. 9.2.

The magnetization of rocks provides then substantial evidence for polar wandering, and for continental drift.

The aspect which has not been investigated sufficiently is the configuration of the earth's magnetic field in the past taking into account the interior of the earth not as it may be now, but as it may have been in times past. This is unlikely to be easy to do. Studies which may lead to this are perhaps those which investigate the growth of the earth's core[246, 149] The contribution of palaeomagnetism to continental drift and polar wandering has been very great. Irving puts the matter well.[148] The hypothesis of polar wandering had been proposed on the basis of geological and other evidence. The study of the magnetization of rocks provided a test of the hypothesis, which it met. Had the directions of magnetization of all rocks of all ages been consistent with the present magnetic field no doubt the matter of polar wandering would have been dropped. Irving also points out that the significant comparison between ancient palaeomagnetic latitudes and climatological indicators which has to be made is with data from the *same* region, and consequently the faunal evidence presented by Stehli, mentioned already, is not a test of palaeomagnetic conclusions.

CONTINENTAL DRIFT: OCEAN-FLOOR SPREADING

The idea that ocean floors may be spreading outwards from the crests of mid-ocean ridges has made continental drift plausible. Evidence that the ocean floors do spread is provided by magnetic field observations over the ridges, interpreted in the way suggested by Vine and Matthews.

The distribution of the world's mid-ocean ridges and mountain ranges is shown in Fig. 9.3. Hess suggested that continental drift could occur if rock was continually brought from the interior of the earth beneath the crests of ridges, to form new oceanic crust. The new crust is transported away from the ridge crests on the backs of convection currents in the mantle, which rise beneath mid-ocean ridges, and turn sideways from the centres. Continents ride upon these currents also. Consequently, if a mid-ocean ridge develops beneath a continent, then the continent will be split, and subsequently the ridge will be located centrally between the now-drifted parts of the continent. Such is the case with the Mid-Atlantic Ridge, and the embryonic rift of the Red Sea will, in time, develop in the same way. If a mid-ocean ridge does not develop beneath a continent, but elsewhere, it will not, except by chance, be located centrally between continental blocks. Such is the East Pacific Rise. In its case, it is being over-ridden by the North American continent, driven by a convection current beneath the Atlantic Ocean.

Material comes up beneath mid-ocean ridges. Where does it again go down? The usual answer given to this question is that it goes down

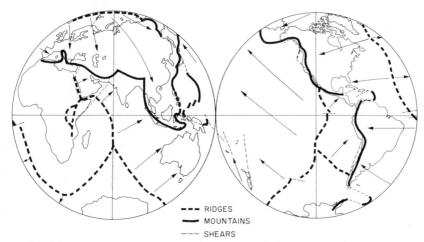

--- RIDGES
▬ MOUNTAINS
--- SHEARS

Fig. 9.3. Mid-ocean ridges, mountain ranges and shear zones (after J. Tuzo Wilson[301]).

beneath deep-sea trenches. Earthquakes beneath the trenches are concentrated on a surface which dips away from the ocean, towards and beneath the continent (see Chapter 8). J. Oliver has suggested that this is caused by the motion of oceanic crust moving down into the mantle. This answer is partly satisfactory. Trenches are found east of the East Pacific Rise off South America, where they are to be expected. They are not found off western North America, because this continent must have ridden over any which existed east of this part of the East Pacific Rise. The Atlantic Ocean and the Indian Ocean are contrary. The only trenches known today in the Atlantic Ocean are in the Caribbean. Perhaps former trenches have been filled by sediment and underlie the continental margins of the Atlantic.

The hypothesis of Vine and Matthews[283, 284] starts from the hypothesis that the ocean floors spread (with accompanying continental drift). It is an attempt to explain the main features of the spatial magnetic anomalies over mid-ocean ridges, described in Chapter 8. The main features of these anomalies are these: they trend parallel to the ridge crests: the central anomaly over the crest is the largest: the anomaly patterns are symmetrical about the crests: and the wavelengths of the anomalies change towards the flanks of the ridges, becoming longer, with the amplitudes increasing.

The new oceanic crust is magnetized in the direction of the earth's field, acquiring a permanent magnetization. The central part of a ridge only contains new crust; the parts of the ridge away from the centre are contaminated by new crust not intruded or extruded centrally, but to one side. If the new crust acquires a magnetization of one polarity, and the crust being contaminated has a magnetization of the other polarity, then the central region (new crust only) will have a larger anomaly associated with it than contaminated regions away from the centre. The crust spreads away from the centre, to each side. Consequently the anomaly pattern generated by the alternation of blocks magnetized normally-reversed-normally-reversed...will be symmetrical about the centre. The ocean-floor now contains a record of reversals of the earth's field. But time-scales of the reversals have been established using lavas on land, and sediments in the ocean basins[339,340] (see Chapter 4), so that dates can be put to each of the normal–reverse boundaries, such as is seen in Fig. 8.10. Now a plot can be made of the distance of the boundary from the ridge crest against the age of the boundary, and the rate of spreading found. The rate is found to lie between one and several centimetres per year. In any one region of a ridge it is fairly constant but the rates vary widely from region to region. For example, the rate of spreading of the East Pacific Rise in the South Pacific, is 4·5 cm/y, but is only 2·9 cm/y over the Juan de Fuca Ridge, west of Vancouver Island.[283]

How does this picture of ocean-floor spreading fit with other character-istics of the ocean basins with which we are familiar? For example, we know that the crests of the mid-ocean ridges are off-set from place to place, and shallow earthquakes are associated with the line of the offset between crests, but not with the line beyond the crests (Fig. 9.4).

J. Tuzo Wilson recognized a difference between wrench or trans-current faults of tradition, and the off-sets of mid-ocean ridges at fracture zones, which he called *transform faults*.[299, 301, 302, 303] In wrench faulting two blocks move relative one to another along a vertical plane (the fault plane). So long as faulting continues, the relative displacement between the two blocks increases. The motion of the block on one side of the fault plane is always opposite in direction from the motion on the other side of the plane. Suppose now that just before a mid-ocean ridge is initiated a wrench fault develops across its future line – a once-for-all displacement (Fig. 9.4). Material brought up from beneath along the line of the ridge (now fractured) is spread sideways in opposite directions

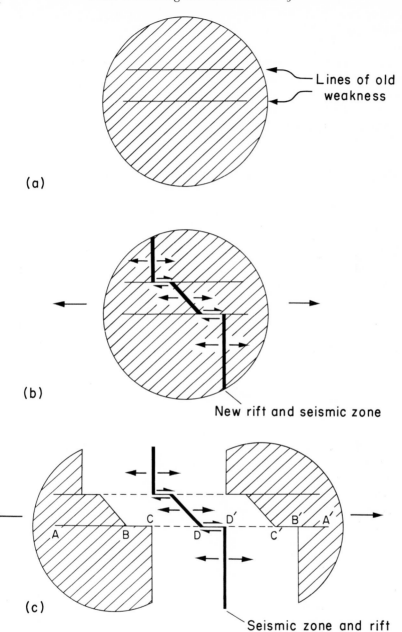

Fig. 9.4. Transform faults (after J. Tuzo Wilson[301]).

between the displaced crest of the ridge. Beyond the displaced crests the spreading material has no relative motion across the now dormant wrench fault. As a consequence, evidence of instability—caused by adjacent blocks moving in opposite directions—will be confined to the zone between the crests, and earthquakes might be expected only in this region, as in fact observed (Chapter 8). This then explains why the fracture zones of the East Pacific are not found beyond the margins of the Rise.

The most direct evidence for ocean-floor spreading comes from studies in Iceland.[25] Walker has estimated that a pile of lava much more than 10 km thick has been extruded here, of which at least 10 km has been deposited since the beginning of the Tertiary. He also points out that the lavas are fed by dykes, and that the total width of dykes intruded is approximately the same as the spreading of Iceland in the same time, 400 km. Note that at the present time it is not thought generally that the intrusion of dykes causes the spreading; it would be difficult for dykes intruded at, say, the centre of the Atlantic Ocean to push Europe and North America aside, without crumpling to occur, not seen. Spreading itself appears to demand convection currents.

Progress in this field could arise not only from studies of the physical properties of the upper mantle of the earth, but also by considering more carefully the implications of Walker's studies of dykes, in conjunction with theoretical and experimental investigations such as Anderson made for ring-dykes and cone-sheets.[4,106,144] If magma is generated at depths within the mantle of some tens of kilometres, then beneath mid-ocean ridges we might expect dyke intrusion which does not reach the surface, and the same spreading action that is observed at the surface in Iceland would apply in the upper parts of the mantle, perhaps to a lesser degree.

An aspect which appears to have been neglected is the relationship between heat-flow and continental motion. If a continent drifts at the rate of 1 cm per year the length of time taken to travel 1000 km is about 100 million years; this time is comparable to the times taken for thermal anomalies encountered at the base of the crust to be noticeable at the surface of the crust. A theory of deformation might be built around such speculation—western North America has encountered the East Pacific Rise and the thermal anomalies cause instability. Such thoughts suggest at least that it would be worthwhile for the change in heat-flow with depth to be measured beneath continents in really long drill-holes. Another field which should be fruitful is the investigations of inhomogeneities in the upper mantle and the relationship between continental crust and mantle. Although on the whole it seems clear that the mantle is inhomogeneous and the mantles beneath oceans and continents differ, the western part of North America appears to sit upon the eastern flank of the East Pacific Rise and it seems unlikely that the same sort of inhomogeneous upper mantle should lie beneath eastern and western North America. Perhaps the inhomogeneities, although very real, are on a regional basis not solely related to the continents which overlie or do not overlie the regions.

Geological Society of London Phanerozoic Time-scale 1964

CAINOZOIC
    Quaternary
        Pleistocene   1·5–2 million years
    Tertiary
        Pliocene   7 (approx.)
        Miocene   26
        Oligocene   37–38
        Eocene   53–54
        Palaeocene   65

MESOZOIC
    Cretaceous   136
    Jurassic   190–195
    Triassic   225

PALAEOZOIC
    Permian   280
    Carboniferous 345
    Devonian   395
    Silurian   430–440
    Ordovician   500 (approx.)
    Cambrian   570

Simplified after Geological Society of London.[95] The ages shown are those of the base of each unit.

Rock-forming Silicates

[Rock-forming minerals are described throughly by Deer, Howie and Zussman.[58]]

1. INDEPENDENT SiO_4 TETRAHEDRA: THE ORTHO-SILICATES

The most important minerals in this group are the *olivines* and *garnets*.

Olivine group

Olivine is the group name for a series of minerals which form the atomic substitution series $(Mg, Fe^{2+})_2 SiO_4$, the end members being *forsterite* Mg_2SiO_4 and *fayalite* $Fe_2^+ SiO_4$. Magnesium-rich olivines are more common than iron-rich ones. Olivines are the principal components of many basic and ultrabasic rocks, such as the gabbros and peridotites.

Garnet Group

Garnet is the group name for a series of minerals with the general formula $R_3^{2+}R_2^{3+}Si_3O_{12}$. (Note that Si_3O_{12} is equivalent formally to $(SiO_4)_3$.) The metal ions R^{2+}, R^{3+}, determine the particular mineral and among the more common ones are

Almandine	$Fe_3^{2+}Al_2Si_3O_{12}$
Grossular	$Ca_3Al_2Si_3O_{12}$
Pyrope	$Mg_3Al_2Si_3O_{12}$

Almandine is the most common in metamorphic and igneous rocks, but grossular and pyrope are abundant in eclogites (a pyroxene–garnet rock) and pyrope in kimberlites (which are ultrabasic diamond-bearing pipes of rock).

Chloritoid

Chloritoid is mica-like superficially and resembles the mineral chlorite. It is in fact one of the group with independent SiO_4 tetrahedra, and has a complex general formula

$$(Fe^{2+}, Mg, Mn)_2 (Al, Fe^{3+}) Al_3O_2 [SiO_4]_2 (OH)_4.$$

It is common in many metamorphic rocks.

2. DOUBLE TETRAHEDRA SILICATES

Melilite Group

The principal member of this group is *melilite*, a calcium-rich mineral

$$(Ca, Na, K)_2 [(Mg, Fe^{2+}, Fe^{3+}, Al, Si)_3 O_7].$$

It occurs as the result of the thermal metamorphism of limestones (i.e., impure limestone is heated up) and in some basalts.

3. RING SILICATES

The most important ring silicates are *beryl* $Be_3Al_2[Si_6O_{18}]$, and *tourmaline*, a rather complicated ring silicate containing hydroxyl groups, boron and fluorine:

$$Na(Mg, Fe, Mn, Li, Al)_3 Al_6 [Si_6O_{18}] (BO_3)_3 (OH, F)_4.$$

4. SINGLE-CHAIN SILICATES: PYROXENES

The pyroxenes are a most important group of minerals. They fall into two groups fairly naturally, defined by their crystallography.

Orthorhombic Pyroxenes: Calcium Free

The orthorhombic pyroxenes, or orthopyroxenes, consist of the atomic substitution series with end members

$$Mg_2Si_2O_6 \qquad \text{enstatite}$$
$$Fe_2^+Si_2O_6 \qquad \text{ferrosilite}$$

In practice the common orthopyroxenes have both magnesium and iron.

Monoclinic Pyroxenes: Calcium Rich or Poor

The monoclinic pyroxenes, or clino-pyroxenes have the general formula

$$(Ca, Mg, Fe^{2+})_2Si_2O_6.$$

Common clino-pyroxenes with all three of calcium, magnesium and iron are *augite* and *diopside*.

5. DOUBLE-CHAIN SILICATES: AMPHIBOLES

Tremolite has the relatively simple formula

$$Ca_2Mg_5[Si_8O_{22}](OH, F)_2.$$

(This illustrates the way that ions of similar size and valency may proxy one for another—here $(OH)^-$, F^- are interchangeable.) The most common amphibole is *hornblende*, which may be described by the formula

$$(Ca, Na, K)_{2-3} (Mg, Fe^{2+}, Fe^{3+}, Al)_5 [Si_6(Si, Al)_2O_{22}] (OH, F)_2.$$

Whilst complicated this does not lack interesting features, for we see again the interchangeability of $(OH)^-$, F^- and see too that Al^{3+} may proxy for Si^{4+} in the $[SiO_4]$ tetrahedron, besides having the role as a coordinating cation. Note that potassium may be present. Hornblende is found in igneous rocks such as basalts and gabbros (among others) and the presence of potassium makes it possible to date the rock by a radioactive method, using the decay of K^{40} to A^{40}. Basalt dykes on the Seychelles Islands have been dated in this way for example.[10]

6 SHEET SILICATES

The sheet silicates include among them the micas and the clay minerals.

Micas

Two are especially important, *muscovite* and *biotite*. Muscovite is colourless and contains potassium, but no magnesium or iron; an ideal muscovite is

$$K_2Al_4 [Si_6Al_2O_{20}] (OH)_4.$$

Biotite is green or brown in colour, and has magnesium and iron as well as potassium. One representative biotite has the formula

$$K_2(Mg, Fe)_6 [Si_6Al_2O_{20}](OH)_4.$$

The presence of potassium makes the minerals suitable for the potassium–argon dating method. They are common in igneous rocks such as granites, less common in rocks such as gabbros.

Chlorites

The chlorites are minerals with layered structures, rather similar to the micas; a general formula is

$$(Mg, Al, Fe)_{12} [(Si, Al)_8O_{20}] (OH)_{16}$$

They are common as alteration products of minerals of igneous rocks, in metamorphic rocks, and in sediments as detrital grains or authigenic (formed in the sediment).

Serpentines

Serpentines are minerals with fibrous habit but sheet structure and include among them the polymorphic forms antigorite, chrysotile and lizardite. A general formula is:

$$Mg_3 (Si_2O_5) (OH)_4.$$

(Chemical analyses quoted by Deer *et al.*[58] show small amounts of Fe^{2+} and other cations which may substitute for magnesium of the general formula.)

Glauconite

Glauconite is a complex mica mineral which can be formed as an authigenic mineral in sediments; most authorities suggest its formation occurs predominantly under marine conditions.[58] For this reason it has been investigated with a view to finding the radioactive age of the sediment in which it occurs.

Clay Minerals

Clay minerals are hydrous silicates, principally of magnesium and aluminium. They can be divided into several groups of which three can be represented by the minerals kaolinite, illite and montmorillonite.

Kaolinite

Kaolinite is formed in the weathering or hydro-thermal alteration of acid rocks, under acidic conditions. Unlike other clay minerals it has a low capacity for cation exchange, because it is quite close to its theoretical formula $Al_4[Si_4O_{10}](OH)_8$. Additional cations may be absorbed at unsatisfied valencies at crystal edges, for example, or by exchange with the hydrogen ions of terminal $(OH)^-$ groups.

Illite

Illite is mica-like and has the general formula

$$K_y Al_4 (Si_{8-y}, Al_y) O_{20} (OH)_4$$

(where y is less than 2, and usually between 1 and 1·5). The cation exchange capacity is less than that of montmorillonite, rather more than that of kaolinite. It occurs as the dominant clay mineral of shales and mudstones and may be derived either by alteration of other silicates, such as feldspars, or from other clay minerals.

Montmorillonite

Montmorillonite is the name for a group of clay minerals of which one has the formula

$$(Na)_{0·7}(Al_{3·3}Mg_{0·7}) Si_8 O_{20} (OH)_4 . n H_2O.$$

These clays have high cation exchange capacity. They form through the alteration of basic igneous rocks, provided sufficient magnesium is present; kaolinite may result otherwise.

7. THREE-DIMENSIONAL FRAMEWORKS: SILICA POLYMORPHS AND FELDSPARS

Silica Polymorphs

These include the abundant mineral *quartz* SiO_2. Quartz is ubiquitous, occurring in many SiO_2-rich igneous rocks and, being resistant to destructive agents is an important constituent of sedimentary rocks. It occurs in two different crystallographic modifications, one (high-quartz) formed at a higher temperature than the other (low-quartz), but which may invert on cooling. It may on cooling retain the external appearance it started with. The observer of the igneous rock may perhaps be able to decide the range in temperatures at which the rock formed. A number of other polymorphs are known, cristobalite, tridymite and coesite among them. These form at higher temperatures, and, in the case of coesite, higher pressures, than does quartz.

Silica which is microcrystalline is found inorganically as flint, for example; the skeletons of organisms such as diatoms, radiolaria and some sponges are also made of silica.

Feldspars

Feldspars are, like pyroxenes, most important rock-forming minerals. They may be divided for convenience into those which contain potassium as the coordinating cation, and those which contain sodium and calcium as the coordinating cation. There is some degree of solid solution between potassium and sodium feldspars, which together form the *alkali* feldspars, and more complete solid solution between sodium and calcium feldspars, which form the *plagioclase* feldspars. The degree of solid solution possible, the structure of the minerals (and the terminology which results) is dependent upon the temperature of formation; there are therefore diagnostic criteria available by which the temperature of formation of the igneous rock in which the feldspars are found may be estimated.

Alkali feldspars

These have the general formula

$$(K, Na) \, AlSi_3O_8$$

and include among them orthoclase (low temperature) and sanidine (high temperature), both predominantly $KAlSi_3O_8$; orthoclase is typically found in granites; sanidine in volcanic rocks such as rhyolites. They resemble SiO_2 in so far as they have a three-dimensional framework, but differ in that one out of every four silicon atoms has been replaced by aluminium.

Plagioclase Feldspars

These form an atomic substitution series with end-members albite, $NaAlSi_3O_8$, and anorthite, $CaAl_2Si_2O_8$. Granites and the *acid* rocks in general have sodium rich plagioclase feldspars, but gabbros have calcium-rich plagioclase feldspars. The names of particular plagioclase feldspars depend upon the relative content of the pure end members:

	$NaAlSi_3O_8$	$CaAl_2Si_2O_8$
	%	%
Albite	100–90	0–10
Oligoclase	90–70	10–30
Andesine	70–50	30–50
Labradorite	50–30	50–70
Bytownite	30–10	70–90
Anorthite	10–0	90–100

Feldspathoids

Feldspathoids resemble the feldspars and silica polymorphs in so far as their structure is a framework. They differ from feldspars in being "deficient" in silica; we can write, for example,

$$NaAlSi_3O_8 - 2SiO_2 = NaAlSiO_4$$
albite $\qquad$ silica $\qquad$ "nepheline"

We expect to find them in alkali rich igneous rocks, which are relatively silica-poor. There are two main groups of importance, the nepheline–kalsilite series, and leucite.

Nepheline–kalsilite

These have end members:

Nepheline	$Na_3K[Al_4Si_4O_{16}]$
Kalsilite	$K[AlSiO_4]$

The degree of solid solution possible depends upon the temperature.

Leucite

This is $KAlSi_2O_6$. Little sodium can be incorporated into the structure. It is a common constituent of potassium-rich, silica-poor igneous rocks.

THE SILICA CONTENT OF MINERALS: SATURATED AND UNDER-SATURATED

We can write the basis units of the mineral groups in the following way:

Structural units		Si	:	O	Example
(1) Individual tetrahedra	SiO_4	4		16	Olivine
(2) Double tetrahedra	Si_2O_7	4		14	Melilite
(3) Ring structures	Si_6O_{18}	4		12	Beryl
(4) Single chains	Si_2O_6				Pyroxenes
(5) Double chains	Si_4O_{11}	4		11	Amphiboles
(6) Sheet structures	Si_4O_{10}	4		10	Micas
(7) Three dimensional framework	SiO_2	4		8	Quartz

We see that the ratio Si:O increases from olivine to quartz; we can in fact formally write equations which express this relationship in a different way, for example,

$$Mg_2SiO_4 + SiO_2 = Mg_2Si_2O_6,$$
$$\text{olivine} + \text{silica} = \text{pyroxene}$$

In a sense then the series of the mineral groups can be considered as being one of continued addition of SiO_2; we say quartz and feldspars are *saturated* with respect to silica, olivines *under-saturated*.

References

1. Allan, T. D., A preliminary magnetic survey in the Red Sea and Gulf of Aden, *Boll. Geofisica Teor. ed. Appl.* **6**, 199–214 (1964).
2. Alldredge, L. R., Keller, F. K. and Dichtel, W. J., Magnetic structure of Bikini Atoll, *U.S. Geol. Surv.* **Prof. Paper 260-L** (1954).
3. Alldredge, L. R., Van Voorhis, G. D. and Davis, T. M., A magnetic profile around the world, *Jour. Geophys. Res.* **68**, 3679–92 (1963).
4. Anderson, E. M., *The Dynamics of Faulting and Dyke Formation with Applications to Britain*, 2nd ed., Oliver & Boyd, Edinburgh (1951).
5. Arnold, J. R., Beryllium-10 produced by cosmic rays, *Science* **124**, 584–5 (1956).
6. Arrhenius, G., Geological record on the ocean floor, in *Oceanography*, 129–48, Ed. M. Sears, Amer. Assoc. Adv. Sci. Washington, D.C. (1961).
7. Arrhenius, G., Pelagic sediments, in *The Sea* **III**, 655–727, Ed. M. N. Hill, Interscience, New York (1963).
8. Backus, G. E., Magnetic anomalies over oceanic ridges, *Nature* **201**, 591–2 (1964).
9. Bagnold, R. A. and Inman, D. L., Beach and nearshore processes, in *The Sea* **III**, 507–53, Ed. M. N. Hill, Interscience, New York (1963).
10. Baker, B. H. and Miller, J. A., Geology and geochronology of the Seychelles Islands and structure of the floor of the Arabian Sea, *Nature* **199**, 346–8 (1963).
11. Barrett, D. L., Berry, M., Blanchard, J. E. Keen, M. J. and McAllister, R. E., Seismic studies on the eastern seaboard of Canada: the Atlantic Coast of Nova Scotia, *Canadian Jour. Earth Sci.* **1**, 10–22 (1964).
12. Barth, T. F. W., *Theoretical Petrology*, 2nd ed., Wiley, New York (1962).
13. Båth, M., Crustal structure of Iceland, *Jour. Geophys. Res.* **65**, 1793–1807 (1960).
14. Benioff, H., Seismic evidence for the fault origin of oceanic deeps, *Bull. Geol. Soc. Amer.* **60**, 1837–56 (1949).
15. Bennington, K. O., Role of shearing stress and pressure in differentiation as illustrated by some mineral reactions in the system $Mgo–SiO_2–H_2O$, *Jour. Geol.* **64**, 558–77 (1956).
16. Berger, J., Blanchard, J. E., Keen, M. J., McAllister, R. E. and Tsong, C. F., Geophysical observations on sediments and basement structure underlying Sable Island, Nova Scotia, *Bull. Amer. Assoc. Petrol. Geol.* **49**, 959–65 (1965).
17. Berry, M. J., and West, G. F., An interpretation of the first-arrival data of the Lake Superior experiment by the time–term method. *Bull. Seismol. Soc. Amer.* **56**, 141–71 (1966).
18. Birch, F., Interpretation of the seismic structure of the crust in the light of experimental studies of wave velocities in rocks, in *Contributions in Geophysics*, 158–70, Ed. H. Benioff, M. Ewing, B. F. Howell and F. Press, Pergamon, London (1958).
19. Birch, F., The velocity of compressional waves in rocks to 10 kilobars, Pt. 1, *Jour. Geophys. Res.* **65**, 1083–102 (1960).
20. Birch, F., The velocity of compressional waves in rocks to 10 kilobars, Pt. 2, *Jour. Geophys. Res.* **66**, 2199–224 (1961).
21. Birch, F., Velocity of compressional waves in serpentinite from Mayaguez, Puerto Rico, in *A study of serpentinite*, 132–3, Ed. C. A. Burk, *Nat. Acad. Sci. – Nat. Res. Council Publ.* **1188**, Washington, D.C. (1964).
22. Biscaye, P. E., Mineralogy and sedimentation of recent deep-sea clay in the Atlantic Ocean and adjacent seas and oceans, *Bull. Geol. Soc. Amer.* **76**, 803–32 (1965).

23. Black, M., Hill, M. N., Laughton, A. S. and Matthews, D. H., Three non-magnetic seamounts off the Iberian coast, *Quart. Jour. Geol. Soc. Lond.* **120**, 477–517 (1964).
24. Blackett, P. M. S., Bullard, E. and Runcorn, S. K. (Eds.), A symposium on continental drift, *Phil. Trans. Roy. Soc. Lond.* A **258**, 1–323 (1965).
25. Bodvarsson, G. and Walker, G. P. L., Crustal drift in Iceland, *Geophys. Jour.* **8**, 285–300 (1964).
26. Bowen, N. L., *The Evolution of the Igneous Rocks*, Dover, New York (1956).
27. Bowen, N. L. and Schairer, J. F., The system $MgO-FeO-SiO_2$, *Amer. Jour. Sci.* **29**, 151–217 (1935).
28. Bowen, N. L. and Tuttle, O. F., The system $MgO-SiO_2-H_2O$, *Bull. Geol. Soc. Amer.* **60**, 439–60 (1949).
29. Braarud, T., Cultivation of marine organisms as a means of understanding environmental influences on populations, in *Oceanography*, 271–98, Ed. M. Sears, Amer. Assoc. Adv. Sci., Washington, D.C. (1961).
30. Bradshaw, J. S., Ecology of living planktonic Foraminifera in the north and equatorial Pacific Ocean, *Contributions Cushman Found. Foraminiferal Research.* **10**, Pt. 2, 25–64 (1959).
31. Bramlette, M. N., Pelagic sediments, in *Oceanography*, 345–66, Ed. M. Sears, Amer. Assoc. Adv. Sci., Washington, D.C. (1961).
32. Brekhovskikh, L. M., *Waves in Lavered Media*, Academic Press, New York (1960).
33. Broecker, W. S., Turekian, K. K. and Heezen, B. C., The relation of deep-sea sedimentation rates to variations in climate, *Amer. Jour. Sci.* **256**, 503–17 (1958).
34. Browne, B. C., The measurement of gravity at sea, *Mon. Not. Roy. Astr. Soc., Geophys. Suppl.* **4**, 271–9 (1937).
35. Brune, J. and Dorman, J., Seismic waves and earth structure in the Canadian shield, *Bull. Seismol. Soc. Amer.* **53**, 167–210 (1963).
36. Buchbinder, G. G. R., Nyland, E. and Blanchard, J. E., Measurement of stress in boreholes, *Upper Mantle Symposium*, Ottawa (1965).
37. Bullard, E., The flow of heat through the floor of the Atlantic Ocean, *Proc. Roy. Soc. Lond.* A **222**, 408–29 (1954).
38. Bullard, E. C., The flow of heat through the floor of the ocean, in *The Sea* **III**, 218–32, Ed. M. N. Hill, Interscience, New York (1963).
39. Bullard, E., Everett, J. E. and Smith, A. G., The fit of the continents around the Atlantic, in *A Symposium on Continental Drift*, Ed. P. M. S. Blackett, E. Bullard and S. K. Runcorn, *Phil. Trans. Roy. Soc. Lond.* A **258**, 41–51 (1965).
40. Bullard, E. C. and Mason, R. G., The magnetic field over the oceans, in *The Sea* **III**, 175–217, Ed. M. N. Hill, Interscience, New York (1963).
41. Bullen, K. E., *An Introduction to the Theory of Seismology*, 2nd ed., Cambridge University Press, Cambridge (1963).
42. Bullen, K. E., Seismic ray theory, *Geophys. Jour.* **4**, 93–105 (1961).
43. Bunce, E. T. Emery, K. O., Gerard, R. D., Knott, S. T., Lidz, L. Saito, T. and Schlee, J., Ocean drilling on the continental margin, *Science* **150**, 709–16 (1965).
44. Burk, C. A. (Ed.), A study of serpentinite, *Nat. Acad. Sci.—Nat. Res. Coun. Publ.* **1188** (1964).
45. Cagniard, L., Electricité tellurique, *Handbuch der Physik*, **47**, 407–69, Springer-Verlag, Berlin (1956).
46. Challis, G. A., The origin of New Zealand ultramatic intrusions, *Jour. Petrol.* **6**, 322–64 (1965).
47. Clark, S. P. and Ringwood, A. E., Density distribution and constitution of the mantle, *Rev. of Geophy.* **2**, 35–88 (1964).
48. Collin, A. E., and Dunbar, M. J., Physical oceanography in Arctic Canada, *Oceanogr. Mar. Biol. Ann. Rev.* **2**, 45–75 (1964).
49. Cook, A. H., Absolute determination of gravity, *Encyclopaedic Dictionary of Physics* **3**, 510–12, Pergamon, London (1961).
50. Cook, A. H., A new absolute determination of the acceleration due to gravity at the National Physical Laboratory, *Nature* **208**, 279 (1965).

51. Cox, A. and Doell, R. R., Review of palaeomagnetism, *Bull. Geol. Soc. Amer.* **71**, 645–768 (1960).
52. Curtis, G. H. and Reynolds, J. H., Notes on the potassium–argon dating of sedimentary rocks, *Bull. Geol. Soc. Amer.* **69**, 151–9 (1958).
53. Daly, R. A., The geology of Ascension Island, *Amer. Acad. Arts and Sci. Proc.* **60**, 1–80 (1925).
54. Darwin, C., *The Structure and Distribution of Coral Reefs*, University of California Press, Berkeley and Los Angeles (1962).
55. Davies, D. and Francis, T. J. G., The crustal structure of the Seychelles Bank, *Deep-Sea Res.* **11**, 921–7 (1964).
56. Davis, W. M., The Coral Reef problem, *Amer. Geograph. Soc. Spec. Pub.* **No. 9**, New York (1928).
57. Deacon, G. E. R., Navigation and the science of the sea, *Jour. Inst. Nav.* **15**, 1–13, (1962).
58. Deer, W. A., Howie, R. A. and Zussman, J., *Rock-Forming Minerals*, **5 V**. Longmans, London (1962).
59. Dobrin, M. B. and Perkins, B., Seismic studies of Bikini Atoll, *U.S. Geol. Surv.* **Prof. Paper 260–J** (1954).
60. Dorman, J., Ewing, M. and Oliver, J., Study of shear-velocity distribution in the upper mantle by mantle Rayleigh waves, *Bull. Seismol. Soc. Amer.* **50**, 87–115 (1960).
61. Drake, C. L., Campbell, N. J., Sander, G. and Nafe, J. E., A Mid-Labrador sea ridge, *Nature* **200**, 1085–6 (1963).
62. Drake, C. L., Ewing, M. and Sutton, G. H., Continental margins and geosynclines: the east coast of North America north of Cape Hatteras, in *Physics and Chemistry of the Earth* **III**, 110–98 (1959).
63. Drake, C. L. and Girdler, R. W., A geophysical study of the Red Sea, *Geophys. Jour.* **8**, 473–95 (1963).
64. Eaton, J. P. and Murata, K. J., How volcanoes grow, *Science* **132**, 925–38 (1960).
65. Emery, K. O., *The Sea off Southern California*, Wiley, New York (1960).
66. Emery, K. O., Tracey, J. I. and Ladd, H. S., Geology of Bikini and nearby atolls, *U. S. Geol. Surv.* **Prof. Paper 260-A**, (1954).
67. Emiliani, C., Pleistocene temperatures, *Jour. Geol.* **63**, 538–78 (1955).
68. Emiliani, C. and Flint, R. F., The Pleistocene record, in *The Sea* **III**, 888–927, Ed. M. N. Hill, Interscience, New York (1963).
69. Engel, A. E. J. and Engel, C. G., Igneous rocks of the East Pacific Rise, *Science*, **146**, 477–85 (1964).
70. Engel, A. E. J. and Engel, C. G., Composition of basalts from the Mid-Atlantic Ridge, *Science* **144**, 1330–3 (1964).
71. Engel, A. E. J., Engel, C. G. and Havens, R. G., Chemical characteristics of oceanic basalts and the upper mantle, *Bull. Geol. Soc. Amer.* **76**, 719–34 (1965).
72. Ericson, D. B., Coiling direction of *Globigerine pachyderma* as a climatic index, *Science* **130**, 219–20 (1959).
73. Ericson, D. B., Ewing, M., Wollin, G. and Heezen, B. C., Atlantic deep-sea sediment cores, *Bull. Geol. Soc. Amer.* **72**, 193–286 (1961).
74. Ericson, D. B. and Wollin, G., Correlation of six cores from the equatorial Atlantic and the Caribbean, *Deep-Sea Res.* **3**, 104–25 (1955/6).
75. Evernden, J. F., Savage, D. E., Curtis, G. H., and James, G. T., Potassium–argon dates and the Cenozoic mammalian chronology of North America, *Amer. Jour. Sci.* **262**, 145–98 (1964).
76. Ewing, G. N., Dainty, A. M., Blanchard, J. E. and Keen, M. J., Seismic studies on the eastern seaboard of Canada: The Appalachian System. I, *Can. Jour. Earth Sci.* **3**, 89–109 (1966).
77. Ewing, J. and Ewing, M., Seismic–refraction measurements in the Atlantic Ocean Basins, in the Mediterranean Sea, on the Mid-Atlantic Ridge, and in the Norwegian Sea, *Bull. Geol. Soc. Amer.* **70**, 291–317 (1959).
78. Ewing, J. and Ewing, M., Reflection-profiling in and around the Puerto Rico trench, *Jour. Geophys. Res.* **67**, 4729–39 (1962).

79. Ewing, J. I. and Nafe, J. E., The unconsolidated sediments, in *The Sea* **III**, 73–84, Ed. M. N. Hill, Interscience, London (1963).
80. Ewing, M., Comments on the theory of glaciation, in *Problems in Palaeoclimatology*, 348–53, Ed. A. E. M. Nairn, Interscience, London (1964).
81. Ewing, M., Ewing, J. I. and Talwani, M., Sediment distribution in the oceans: the Mid-Atlantic Ridge, *Bull. Geol. Soc. Amer.* **75**, 17–36 (1964).
82. Ewing, M., Jardetzky, W. S. and Press, F., *Elastic Waves in Layered Media*, McGraw-Hill, New York (1957).
83. Ewing, M. and Landisman, M., Shape and structure of ocean basins, in *Oceanography*, 3–38, Ed. M. Sears, Amer. Assoc. Adv. Sci. Washington, D.C. (1961).
84. Faure, G. and Hurley, P. M., The isotopic composition of strontium in oceanic and continental basalts: application to the origin of igneous rocks, *Jour. Petrol.* **4**, 31–50 (1963).
85. Findlay, D. C. and Smith, C. H., The Muskox drilling project, *Geol. Sur. Can.* **Pap. 64–44** (1964).
86. Fisher, R. L. and Hess, H. H., Trenches, in *The Sea* **III**, 411–36, Ed. M. N. Hill, Interscience, New York (1963).
87. Fisher, R. L. and Raitt, R. W., Topography and structure of the Peru–Chile trench, *Deep-Sea Res.* **9**, 423–43 (1962).
88. Francis, T. J. G., Black Mud Canyon, *Deep-Sea Res.* **9**, 457–64 (1962).
89. Fredriksson, K. and Martin, L. R., The origin of black spherules found in Pacific Islands, deep-sea sediments and Antarctic ice, *Geochim. et Cosmochim. Acta.* **27**, 245–8 (1963).
90. Fuller, M. D., Expression of E–W fractures in magnetic surveys in parts of the U.S.A., *Geophysics* **29**, 602–22 (1964).
91. Futterman, W. I., Dispersive body waves, *Jour. Geophys. Res.* **67**, 5279–91 (1962).
92. Garland, G. D., *The Earth's Shape and Gravity*, Pergamon, London (1965).
93. Gast, P. W., Limitations on the composition of the upper mantle, *Jour. Geophys. Res.* **65**, 1287–97 (1960).
94. Gast, P. W., Tilton, G. R. and Hedge, C., Isotopic composition of lead and strontium from Ascension and Gough Islands, *Science* **145**, 1181–5 (1964).
95. Geological Society Phanerozoic time-scale 1964, *Quart. Jour. Geol. Soc. Lond.* **120**, 260–2 (1964).
96. Gignoux, M., *Stratigraphic Geology*, Freeman, San Francisco (1955).
97. Girdler, R. W., Geophysical studies of rift valleys, *Physics and Chemistry of the Earth*, **5**, 120–56 (1964).
98. Goldberg, E. D. and Arrhenius, G. O. S., Chemistry of Pacific pelagic sediments, *Geochim. et Cosmochim. Acta* **13**, 153–212 (1958).
99. Goldberg, E. D. and Griffin, J. J., Sedimentation rates and mineralogy in the South Atlantic, *Jour. Geophys. Res.* **69**, 4293–309 (1964).
100. Goldberg, E. D. and Koide, M., Geochronological studies of deep-sea sediments by the ionium/thorium method, *Geochim. et Cosmochim.. Acta* **26**, 417–50 (1962).
101. Goldberg, E. D., Koide, M., Griffin, J. J. and Peterson, M. N. A. A geochronological and sedimentary profile across the North Atlantic Ocean, in *Isotopic and Cosmic Chemistry*, 211–32, Ed. H. Craig, S. L. Miller and G. J. Wasserburg, North-Holland, Amsterdam (1964).
102. Grant, F. S., A problem in the analysis of geophysical data, *Geophysics* **22**, 309–44 (1957).
103. Griffin, J. J. and Goldberg, E. D., Clay-mineral distributions in the Pacific Ocean, in *The Sea* **III**, 728–41, Ed. M. N. Hill, Interscience, New York (1963).
104. Gutenberg, B., On the layer of relatively low wave velocity, at a depth of about 80 kilometers, *Bull. Seismol. Soc. Amer.* **38**, 121–48 (1948).
105. Gutenberg, B. and Richter, C. F., *Seismicity of the Earth and Related Phenomena*, 2nd ed. Princeton University Press, Princeton (1954).
106. Hafner, W., Stress distributions and faulting, *Geol. Soc. Amer. Bull.* **62**, 373–98 (1951).
107. Hamilton, E. L., Sunken islands of the Mid-Pacific mountains, *Geol. Soc. Amer.* **Mem. 64** New York (1956).
108. Hamilton, E. L., Marine geology of the southern Hawaiian Ridge, *Bull. Geol. Soc. Amer.* **68**, 1011–26 (1957).

109. Hamilton, E. L., Thickness and consolidation of deep-sea sediments, *Bull. Geol. Soc. Amer.* **70**, 1399–1424 (1959).
110. Harland, W. B., Smith, A. G. and Wilcock, B. (Eds.), The Phanerozoic time-scale, *Quart. Jour. Geol. Soc. Land.* **120s** (1964).
111. Hawkes, L., Icelandic tectonics—graben or horst? *Geol. Mag.* **78**, 305–8 (1941).
112. Hayes, F. R., The mud-water interface, *Oceanogr. Mar. Biol. Ann. Rev.* **2**, 121–45 (1964).
113. Hedge, C. E. and Walthall, F. G., Radiogenic strontium-87 as an index of geologic processes, *Science* **140**, 1214–17 (1963).
114. Heezen, B. C., Dynamic processes of abyssal sedimentation: erosion, transportation and redeposition on the deep-sea floor, *Geophys. Jour.* **2**, 142–63 (1959).
115. Heezen, B. C., Turbidity currents, in *The Sea* **III**, Ed. M. N. Hill, 742–75, Interscience, New York (1963).
116. Heezen, B. C., Bunce, E. T., Hersey, J. B. and Tharp, M., Chain and Romanche fracture zones, *Deep-Sea Res.* **11**, 11–33 (1964).
117. Heezen, B. C. and Drake, C. L., Grand Banks slump, *Amer. Assoc. Petrol. Geol. Bull.* **48**, 221–5 (1964).
118. Heezen, B. C., Ericson, D. B. and Ewing, M., Further evidence for a turbidity current following the 1929 Grand Banks earthquake, *Deep-Sea Res.* **1**, 193–202 (1953/4).
119. Heezen, B. C. and Ewing, M., Turbidity currents and submarine slumps, and the 1929 Grand Banks earthquake, *Amer. Jour. Sci.* **250**, 849–73 (1952).
120. Heezen, B. C. and Ewing, M., The mid-oceanic ridge and its extension through the Arctic Basin, in *Geology of the Arctic* **1**, 622–42, Ed. G. O. Raasch, University of Toronto Press, Toronto (1961).
121. Heezen, B. C. and Ewing, M., The mid-oceanic ridge, in *The Sea* **III**, 388–410, Ed. M. N. Hill, Interscience, New York (1963).
122. Heezen, B. C., Ewing, M. and Miller, E. T., Trans-Atlantic profile of total magnetic intensity and topography, Dakar to Barbados, *Deep-Sea Res.* **1**, 25–33 (1953/4).
123. Heezen, B. C., Gerard, R. D. and Tharp, M., The Vema fracture zone in the equatorial Atlantic, *Jour. Geophys. Res.* **69**, 733–9 (1964).
124. Heezen, B. C. and Hollister, C., Deep-sea current evidence from abyssal sediments, *Mar. Geol.* **1**, 141–74 (1964).
125. Heezen, B. C. and Laughton, A. S., Abyssal plains, in *The Sea* **III**, 312–64, Ed. M. N. Hill, Interscience, New York (1963).
126. Heezen, B. C. and Menard, H. W., Topography of the deep-sea floor, in *The Sea* **III**, 233–80, Ed. M. N. Hill, Interscience, New York (1963).
127. Heezen, B. C. and Nafe, J. E., Vema Trench: western Indian Ocean, *Deep-Sea Res.* **11**, 79–84 (1964).
128. Heezen, B. C. and Tharp, M., *Physiographic Diagram of the South Atlantic, the Caribbean Sea, the Scotia Sea, and the Eastern Margin of the South Pacific Ocean*, Geol. Soc. Amer., New York (1962).
129. Heezen, B. C., Tharp, M. and Ewing, M., The floors of the oceans, **I**. The North Atlantic, *Geol. Soc. Amer. Spec. Pap.* **65**, New York (1959).
130. Heirtzler, J. R. and Le Pichon, X., Crustal structure of the mid-ocean ridges. 3. Magnetic anomalies over the mid-Atlantic ridge, *Jour. Geophys. Res.* **70**, 4013–33 (1965).
131. Hersey, J. B., Findings made during the June 1961 cruise of *Chain* to the Puerto Rico trench and Caryn seamount, *Jour. Geophys. Res.* **67**, 1109–16 (1962).
132. Herzen, R. P. Von, Heat-flow values from the south-eastern Pacific, *Nature* **183**, 882–3 (1959).
133. Hess, H. H., History of ocean basins, in *Petrologic Studies: a Volume in Honour of A. F. Buddington*, 599–620, Ed. A. E. J. Engel and others, Geol. Soc. Amer., New York (1962).
134. Hess, H. H., The oceanic crust, the upper mantle and serpentinized peridotite, in *A Study of Serpentinite*, Ed. C. A. Burk, Nat. Acad. Sci.—Nat. Res. Coun. **Publ. 1188** 169–175, Washington, D.C. (1964).
135. Hill, M. L. and Dibblee, T. W., San Andreas, Garlock, and Big Pine faults, California, *Bull. Geol. Soc. Amer.* **64**, 443–58 (1953).
136. Hill, M. N., Recent geophysical exploration of the ocean floor, *Physics and Chemistry of the Earth* **2**, 129–63 (1957).

137. Hill, M. N., A median valley of the Mid-Atlantic Ridge, *Deep-Sea Res.* **6**, 193–205 (1960).
138. Hill, M. N. and Mason, C. S., Diurnal variation of the earth's magnetic field at sea, *Nature* **195**, 365–6 (1962).
139. Hill, M. N. and Vine, F. J., A preliminary magnetic survey of the Western Approaches to the English Channel, *Quart. Jour. Geol. Soc. Lond.* **121**, 463–75 (1965).
140. Hodgson, J. H., Nature of faulting in large earthquakes, *Bull. Geol. Soc. Amer.* **68**, 611–43 (1957).
141. Holmes, A., *Principles of Physical Geology*, New ed., Nelson, London (1965).
142. Holtedahl, O. and Holtedahl, H., On "marginal channels" along continental borders and the problem of their origin, *Bull. Geol. Inst. Univ. Uppsala* **40**, 183–7 (1961).
143. Hospers, J., Reversals of the main geomagnetic field, *K. Nederl. Akad. Wetens., Pr.*, **56**, 467–91 (1953).
144. Hubbert, M. K., Mechanical basis for certain familiar geologic structures, *Geol. Soc. Amer. Bull.* **62**, 355–72 (1951).
145. Hunter, W. and Parkin, D. W., Cosmic dust in recent deep-sea sediments, *Proc. Roy. Soc. Lond. A.* **255**, 382–97 (1960).
146. Hurley, P. M., Heezen, B. C., Pinson, W. H. and Fairbairn, H. W., K–Ar age values in pelagic sediments of the North Atlantic, *Geochim. et Cosmochim. Acta* **27**, 393–9 (1963).
147. Hutchinson, R. D., Cambrian stratigraphy and trilobite faunas of southeastern Newfoundland, *Bull. Geol. Surv. Can.* **88** (1962).
148. Irving, E., *Palaeomagnetism and its Application to Geological and Geophysical Problems*, Wiley, New York (1964).
149. Jacobs, J. A., *The Earth's Core and Geomagnetism*, Pergamon, London (1963).
150. Jacobs, J. A. and Westphal, K. O., Geomagnetic micropulsations, *Physics and Chemistry of the Earth* **5**, 157–224 (1964).
151. Johnson, M. A., Turbidity currents, *Oceanogr. Mar. Biol. Ann. Rev.* **2**, 31–43 (1964).
152. Johnson, M. W. and Brinton, E., Biological species, water masses and currents, in *The Sea* **II**, 381–414, Ed. M. N. Hill, Interscience, New York (1963).
153. Jousé, A. P., Les diatomées des dépôts de fond de la partie nord-ouest de l'Océan Pacifique, *Deep-Sea Res.* **6**, 187–92 (1959/60).
154. Juignet, P., Dangeard, L. and Le Guyader, M.-T., Les courants de turbidité et les turbidites, *Rev. Geograph. Phys. et de Geol. Dynam.*, 2nd ser., **7**, 97–122 (1965).
155. Kaplan, I. R., Emery, K. O. and Rittenberg, S. C., The distribution and isotopic abundance of sulphur in recent marine sediments off southern California, *Geochim. et Cosmochim. Acta* **27**, 297–331 (1963).
156. Keen, C. E. and Loncarevic, B. D., Crustal structure on the eastern seaboard of Canada: studies on the continental margin, *Can. Jour. Earth Sci.* **3**, 65–76 (1966).
157. Keen, M. J., Magnetic anomalies over the Mid-Atlantic Ridge, *Nature* **197**, 888–90 (1963).
158. Keen, M. J., The magnetization of sediment cores from the eastern basin of the North Atlantic Ocean, *Deep-Sea Res.* **10**, 607–22 (1963).
159. Kennedy, W. Q., The Great Glen Fault, *Quart. Jour. Geol. Soc. Lond.* **102**, 41–76 (1946).
160. Koczy, F. F., Age determination in sediments by natural radioactivity, in *The Sea* **III**, 816–31, Ed. M. N. Hill, Interscience, New York (1963).
161. Kuenen, P. H., *Marine Geology*, Wiley, New York (1950).
162. Kuenen, P. H., Topographie et géologie des profondeurs océaniques, in *Problèmes de géologie sous-marine*, Ed. J. Bourcart, Paris, Masson (1958).
163. Kuenen, P. H. and Carozzi, A., Turbidity currents and sliding in geosynclinal basins of the Alps, *Jour. Geol.* **61**, 363–73 (1953).
164. Kulp, J. L., Geologic time-scale, *Science* **133**, 1105–14 (1961).
165. Lambert, A. and Caner, B., Geomagnetic "depth-sounding" and the coast effect in Western Canada, *Can. Jour. Earth Sci.* **2**, 485–509 (1965).
166. Laughton, A. S., Laboratory measurements of seismic velocities in ocean sediments, *Proc. Roy. Soc. Lond. A*, **222**, 336–41 (1954).

167. Laughton, A. S., An interplain deep-sea channel system, *Deep-Sea Res.* **7**, 75–88 (1960).

168. Le Pichon, X., Houtz, R. E., Drake, C. L. and Nafe, J. E., Crustal structure of the mid-ocean ridges. 1. Seismic refraction measurements, *Jour. Geophys. Res.* **70**, 319–39 (1965).

169. Lee, W. H. K. (Ed.) *Terrestrial Heat-Flow*, Amer. Geophys. Union Monograph No. 8 Washington, D.C. (1965).

170. Lehmann, I., Seismic time-curves and depth determination, *Mon. Not. Roy. Astr. Soc., Geophys. Suppl.* **4**, 250–71 (1937/9).

171. Lipson, J., Potassium–argon dating of sedimentary rocks, *Bull. Geol. Soc. Amer.* **69**, 137–49 (1958).

172. Lochman-Balk, C. and Wilson, J. L., Cambrian biostratigraphy in North America, *Jour. Palaeont.* **32**, 312–50 (1958).

173. Loncarevic, B. D., Measurement of gravity at sea, *Encyclopaedic Dictionary of Physics* **3**, 513–15, Pergamon, London (1961).

174. Loncarevic, B. D., Measurement of gravity on land, *Encyclopaedic Dictionary of Physics* **3**, 515–19, Pergamon, London (1961).

175. Loncarevic, B. D., Geophysical studies in the Indian Ocean, *Endeavour* **23**, 43–47 (1964).

176. Loncarevic, B. D., Bouguer anomaly of an area of Mid-Atlantic Ridge, *Bull. d'Inf. Bur. Grav. Int.* **No. 12** (1966).

177. Loncarevic, B. D., Mason, C. S. and Matthews, D. H., Mid-Atlantic Ridge near 45°N: 1. The median valley (in press).

178. Loncarevic, B. D. and Matthews, D. H., Bathymetric, magnetic and gravity investigations, H.M.S. Owen 1961–1962, *Admiralty Marine Science Publications* **No. 4, Pts. 1–2**, Admiralty, London (1963).

179. Lotze, F., The distribution of evaporites in space and time, in *Problems in Palaeoclimatology*, 491–507, Ed., A. E. M. Nairn, Interscience, London (1964).

180. Lovering, J. F., The nature of the Mohorovičić discontinuity, *Trans. Amer. Geophys. Union* **39**, 947–55 (1958).

181. Lowenstam, H. A., Mineralogy, O^{18}/O^{16} ratios, and strontium and magnesium contents of recent and fossil brachiopods and their bearing on the history of the oceans, *Jour. Geol.* **69**, 241–60 (1961).

182. MacDonald, G. A., Volcanology, *Science* **133**, 673–9 (1961).

183. MacDonald, G. A. and Katsura, T., Variations in the lava of the 1959 eruption in Kilauea Iki, *Pacific Sci.* **15**, 358–69 (1961).

184. MacDonald, G. A. and Katsura, T., Chemical composition of Hawaiian lavas, *Jour. Petrol.* **5**, 82–133 (1964).

185. MacDonald, G. J. F., Geophysical deductions from observations of heat-flow, in *Terrestrial Heat Flow*, Ed. W. H. K. Lee, Amer. Geophys. Union Monograph No. 8, Washington, D.C. 191–210 (1965).

186. MacDonald, G. J. F., The deep structure of continents, *Rev. of Geophys.* **1**, 587–665 (1963).

187. Manchester, K. S., McGrath, P. and Keen, M. J., Geophysical studies in the Labrador Sea, Davis Strait and Baffin Bay, *Trans. Amer. Geophys. Union* **45**, 72 (1964).

188. Marlowe, J. I., Probable Tertiary sediments from a submarine canyon off Nova Scotia, *Mar. Geol.* **3**, 263–8 (1965).

189. Martin, H., The hypothesis of continental drift in the light of recent advances of geological knowledge in Brazil and in South West Africa, *Trans. Geol. Soc. South Africa, B* **Annexure to volume 64**, 1–47 (1961).

190. Mason, R. G. and Raff, A. D., Magnetic survey off the west coast of North America, 32°N. latitude to 42°N. latitude, *Bull. Geol. Soc. Amer.* **72**, 1259–65 (1961).

191. Matthews, D. H., Lavas from an abyssal hill on the floor of the North Atlantic Ocean, *Nature* **190**, 158–9 (1961).

192. Matthews, D. H., Altered lavas from the floor of the eastern North Atlantic, *Nature* **194**, 368–9 (1962).

193. Matthews, D. H., A major fault scarp under the Arabian Sea displacing the Carlsberg Ridge near Socotra, *Nature* **198**, 950–2 (1963).
194. Matthews, D. H., Mid-ocean ridges, in *Dictionary of Geophysics*, Ed. S. K. Runcorn, Pergamon, London (in press).
195. Matthews, D. H., Vine, F. J. and Cann, J. R., Geology of an area of the Carlsberg Ridge, Indian Ocean, *Bull. Geol. Soc. Amer.* **76**, 675–82 (1965).
196. Matthews, D. J., *Tables of the Velocity of Sound in Pure Water and Sea Water for Use in Echo-sounding and Sound-ranging*, 2nd ed. Gt. Brit. Hydrographic Office. **Pub. 282**, London, (1939).
197. Medioli, F., Stanley, D. J. and James, N., The physical influence of palaeosol on the morphology and preservation of Sable Island off the coast of Nova Scotia, *VII Congress of Internat. Quat. Res. Ass. Boulder, Col.* (1965).
198. Menard, H. W., The East Pacific Rise, *Science* **132**, 1737–46 (1960).
199. Menard, H. W., *Marine Geology of the Pacific*, McGraw-Hill, New York (1964).
200. Menard, H. W., Correlation between Length and Offset on Very Large Wrench Faults, *Jour. Geophys. Res.* **67**, 4096–98 (1962).
201. Menard, H. W., The world-wide oceanic rise-ridge system, in *A symposium on continental drift*, Ed. P. M. S. Blackett, E. Bullard and S. K. Runcorn, *Phil. Trans. Roy. Soc. Lond.* A **258**, 109–22 (1965).
202. Menard, H. W., Chase, T. E. and Smith, S. M., Galapagos Rise in the southeastern Pacific, *Deep-Sea Res.* **11**, 233–42 (1964).
203. Menard, H. W. and Fisher, R. L., Clipperton fracture zone in the northeastern equatorial Pacific, *Jour. Geol.* **66**, 239–53 (1958).
204. Menard, H. W. and Ladd, H. S., Oceanic islands, seamounts, guyots and atolls, in *The Sea* **3**, Ed. M. N. Hill, 365–87, Interscience, New York (1963).
205. Miller, J. A. and Mudie, J. D., Potassium–argon age determination on granite from the island of Mahé in the Seychelles Archipelago, *Nature* **192**, 1174–5 (1961).
206. Moorbath, S. and Bell, J. D., Strontium-isotope abundance studies and rubidium–strontium age determinations on Tertiary igneous rocks from the Isle of Skye, north-west Scotland, *Jour. Petrol.* **6**, 37–66 (1965).
207. Moore, D. G. and Curray, J. R., Structural framework of the continental terrace, northwest Gulf of Mexico, *Jour. Geophys. Res.* **68**, 1725–47 (1963).
208. Moore, R. C. (Ed.), *Treatise on Invertebrate Palaeontology. Part F. Coelenterata*, Geological Society of America and University of Kansas Press, New York and Lawrence (1956).
209. Muir, I. D., Tilley, C. E. and Scoon, J. H., Basalts from the northern part of the rift zone of the Mid-Atlantic Ridge, *Jour. Petrol.* **5**, 409–34 (1964).
210. Munk, W. H. and Davies, D., The relationship between core accretion and the rotation rate of the earth, in *Isotopic and Cosmic Chemistry*, 341–6, Ed., H. Craig, S. L. Miller and G. J. Wasserburg, North-Holland, Amsterdam (1964).
211. Nafe, J. E. and Drake, C. L., Physical properties of marine sediments, in *The Sea* **III**, 794–815, Ed. M. N. Hill, Interscience, New York (1963).
212. Nairn, A. E. M. (Ed), *Problems in Palaeoclimatology*, Interscience, London (1964).
213. Nakamura, Y., Model experiments on refraction arrivals from a linear transition layer, *Bull. Seismol. Soc. Amer.* **54**, 1–8 (1964).
214. Nakamura, Y. and Howell, B. F., Maine seismic experiment: frequency spectra of refraction arrivals and the nature of the Mohorovičić discontinuity, *Bull. Seismol. Soc. Amer.* **54**, 9–18 (1964).
215. Neaverson, E., *Stratigraphical Palaeontology*, 2nd ed. Clarendon Press, Oxford (1955).
216. Nesteroff, W. D. and Heezen, B. C., Essais de comparaison entre les turbidites modernes et le flysch, *Rev. Géograph. Phys. Géol. Dynam.*, 2nd ser., **5**, 113–25 (1962).
217. Nicholls, G. D., Nalwalk, A. J. and Hays, E. E., The nature and composition of rock samples dredged from the Mid-Atlantic Ridge between 22°N. and 52°N., *Marine Geol.* **1**, 333–43 (1964).
218. Nuttli, O., Seismological evidence pertaining to the structure of the earth's upper mantle, *Rev. of Geophys.* **1**, 351–400 (1963).

219. Officer, C. B., *Introduction to the Theory of Sound Transmission with Application to the Ocean*, McGraw-Hill, New York (1958).
220. Officer, C. B., The refraction arrival in water-covered areas, *Geophysics* **18**, 805–19 (1953).
221. O'Hara, M. J. and Mercey, E. L. P., Petrology of some garnetiferous peridotites, *Trans. Roy. Soc. Edin.* **65**, 251–314 (1963).
222. Peterson, M. N. A. and Goldberg, E. D., Feldspar distributions in South Pacific pelagic sediments, *Jour. Geophys. Res.* **67**, 3477–92 (1962).
223. Phleger, F. B., *Ecology and Distribution of Recent Foraminifera*, Johns Hopkins Press, Baltimore (1960).
224. Phleger, F. B., Foraminiferal ecology and marine geology, *Marine Geol.* **1**, 16–43 (1964).
225. Picciotto, E. E., Geochemistry of radioactive elements in the ocean and the chronology of deep-sea sediments, in *Oceanography*, 367–90, Ed. M. Sears, Amer. Assoc. Adv. Sci., Washington, D. C. (1961).
226. Powell, J. L., Faure, G. and Hurley, P. M., Strontium-87 abundance in a suite of Hawaiian volcanic rocks of varying silica content, *Jour. Geophys. Res.* **70**, 1509–13 (1965).
227. Pratt, R. M., Great Meteor Seamount, *Deep-Sea Res.* **10**, 17–25 (1963).
228. Pratt, R. M. and Heezen, B. C., Topography of the Blake Plateau, *Deep-Sea Res.* **11**, 721–8 (1964).
229. Quon, S. H. and Ehlers, E. G., Rocks of northern part of Mid-Atlantic Ridge, *Bull. Geol. Soc. Amer.* **74**, 1–8 (1963).
230. Raff, A. D. and Mason, R. G., Magnetic survey off the west coast of North America, 40°N. latitude to 52°N. latitude, *Bull. Geol. Soc. Amer.* **72**, 1267–70 (1961).
231. Raitt, R. W., Seismic refraction studies of Bikini and Kwajalein atolls and Sylvania guyot. *U. S. Geol. Surv.* **Prof. Paper 260-K** (1954).
232. Raitt, R. W., Seismic refraction studies of Eniwetok Atoll, *U. S. Geol. Surv.* **Prof. Paper 260-S** (1957).
233. Raitt, R. W., The crustal rocks, in *The Sea* **III**, 85–102, Ed. M. N. Hill, Interscience, New York (1963).
234. Rankama, K., *Progress in Isotope Geology*, Interscience, New York (1963).
235. Rees, A. I., Measurements of the natural remanent magnetism and anisotropy of susceptibility of some Swedish glacial silts, *Geophys. Jour.* **8**, 356–69 (1964).
236. Revelle, R. R., Marin bottom samples collected in the Pacific Ocean by the Carnegie on its seventh cruise, Carnegie Inst. Wash. Publ. **556**, *Oceanography II*, Pt. 1 (1944).
237. Revelle, R. R. D. and Maxwell, A. E., Heat-flow through the floor of the eastern North Pacific Ocean, *Nature* **170**, 199–200 (1952).
238. Rex, R. W. and Goldberg, E. D., Quartz contents of pelagic sediments of the Pacific Ocean, *Tellus* **10**, 153–9 (1958).
239. Riedel, W. R., The preserved record: paleontology of pelagic sediments, in *The Sea* **3**, 866–87, Ed. M. N. Hill, Interscience, New York (1963).
240. Riley, G. A., Wangersky, P. J. and Van Hemert, D., Organic aggregates in tropical and subtropical surface waters of the North Atlantic Ocean, *Limn. and Oceanog.* **9**, 546–50 (1964).
241. Rona, E., Geochronology of marine and fluvial sediments, *Science*, **144**, 1595–7 (1964).
242. Rosholt, J. N., Emiliani, C., Geiss, J., Koczy, F. F., and Wangersky, P. J., Absolute dating of deep-sea cores by the Pa^{231}/Th^{230} method, *Jour. Geol.* **69**, 162–85 (1961).
243. Ross, D. A. and Shor, G. G., Reflection profiles across the Middle America trench, *Jour. Geophys. Res.* **70**, 5551–72 (1965).
244. Rothé, J. P., La zone séismique mediane Indo-Atlantique, *Proc. Roy. Soc. Lond.* A, **222**, 387–97 (1954).
245. Runcorn, S. K. (Ed.), *Continental Drift*, Academic Press, New York (1962).
246. Runcorn, S. K., A growing core and a convecting mantle, in *Isotopic and Cosmic Chemistry*, 321–40, Ed. H. Craig, S. L. Miller and G. J. Wasserburg, North-Holland, Amsterdam (1964).
247. Runcorn, S. K., Palaeomagnetic comparisons between Europe and North America,

in *A Symposium on Continental Drift*, Ed. P. M. S. Blackett, E. Bullard, and S. K. Runcorn. *Phil. Trans. Roy. Soc. Lond.* A, **258**, 1–11 (1965).

248. Russell, R. D. and Allan, D. W., The age of the earth from lead isotope abundances, *Mon. Not. Roy. Astron. Soc. Geophys. Suppl.* **7**, 80–101 (1955).

249. Schott, W., 1. Die Sedimenté des äquatorialen Atlanuscher Ozeans; 1. Lieferung, B., Die foraminiferen in dem äquatorialen Teil des Atlantischen Ozeans, *Wiss. Ergen. Deutsch. Atlantische Exped. METEOR. 1925–1927* **3** (3), 43–134 (1935).

250. Serson, P. H., Mack, S. Z. and Whitham, K. A three-component airborne magnetometer, *Pub. Dom. Obs. Ottawa* **19**, No. 2, (1957).

251. Shor, G. G., Jr., Structure of the Bering Sea and the Aleutian Ridge, *Mar. Geol.* **1**, 213–219 (1964).

252. Shor, G. G., Jr. and Raitt, R. W., Seismic studies in the Southern California Continental borderland, *Internat. Geol. Cong. 20th*, Mexico, 1956, **sec. 9**, 243–59 (1958).

253. Sillén, L. G., The physical chemistry of sea water, in *Oceanography* 549–81, Ed. M. Sears, Amer. Assoc. Adv. Sci., Washington, D. C. (1961).

254. Stacey, F. D., Dielectric anisotropy and fabric of rocks, *Geofisica Pura e Applicata* **48**, 40–48 (1961).

255. Stanley, D. J., Vertical petrographic variability in Annot Sandstone turbidites: some preliminary observations and generalizations, *Jour. Sed. Pet.* **33**, 783–8 (1963).

256. Stearns, H. T., Geology of the Hawaiian Islands, *Hawaii Div. of Hydrography, Bull.* **8**, (1946).

257. Stehli, F. G., Permian zoogeography and its bearing on climate, in *Problems in Palaeoclimatology*, 537–49, Ed., A. E. M. Nairn, Interscience, London (1964).

258. Steinhart, J. S., and others, The earth's crust: seismic studies, Year Book 61, *Carnegie Institution of Washington*, 221–34 (1962).

259. Steinhart, J. S. and Meyer, R. P., Explosion studies of continental structure, *Carnegie Institution Washington* **Publication 622**, Washington, D.C. (1961).

260. Stockwell, C. H. and Williams, H., Age determinations and geological studies, *Geol. Surv. Can.* **Paper 64–17 (Part II)** (1964).

261. Sverdrup, H. U., Johnson, M. W. and Fleming, R. H., *The Oceans: Their Physics, Chemistry, and General Biology*, Prentice-Hall, New York (1942).

262. Sykes, L. R., Seismicity of the South Pacific Ocean, *Jour. Geophys. Res.* **68**, 5999–6006 (1963).

263. Sykes, L. R., The seismicity of the Arctic, *Bull. Seismol. Soc. Amer.* **55**, 501–18 (1965).

264. Sykes, L. R. and Ewing, M., The seismicity of the Caribbean region, *Jour. Geophys. Res.* **70**, 5065–74 (1965).

265. Sykes, L. R. and Landisman, M., The seismicity of East Africa, the Gulf of Aden and the Arabian and Red Seas, *Bull. Seismol. Soc. Amer.* **54**, 1927–40 (1964).

266. Takeuchi, H., Saito, M. and Kobayashi, N., Study of shear velocity distribution in the upper mantle by mantle Rayleigh and Love waves, *Jour. Geophys. Res.* **67**, 2831–39 (1962).

267. Talwani, M., A review of marine geophysics, *Mar. Geol.* **2**, 29–80 (1964).

268. Talwani, M., Le Pichon, X. and Heirtzler, J. R., East Pacific Rise: the magnetic pattern and the fracture zones, *Science* **150**, 1109–15 (1965).

269. Talwani, M., Le Pichon, X., and Ewing, M., Crustal structure of the mid-ocean ridges. 2. Computed model from gravity and seismic refraction data, *Jour. Geophys. Res.* **70**, 341–52 (1965).

270. Talwani, M., Sutton, G. H. and Worzel, J. L., A crustal section across the Puerto Rico Trench, *Jour. Geophys. Res.* **64**, 1545–55 (1959).

271. Teichert, C., Cold- and deep-water coral banks, *Bull. Amer. Assoc. Petrol. Geol.* **42**, 1064–82 (1958).

272. Thorarinsson, S., On the geology and geomorphology of Iceland, Guide to excursion no. A2. Intern. Geol. Cong. 19th, Stockholm, Excursion guidebook E.I.1. 1960, *Mus. of Nat. Hist., Dept. of Geol. and Geog., Reykjavik*, **Misc. Papers 25** (1960).

273. Thorson, G., Length of pelagic larval life in marine bottom invetebrates as related to larval transport by ocean currents, in *Oceanography*, 455–74, Ed. M. Sears, Amer. Assoc. Adv. Sci., Washington, D.C. (1961).

274. Tilley, C. E., The dunite-mylonites of St. Paul's Rocks (Atlantic), *Amer. Jour. Sci.* **245**, 483–91 (1947).
275. Tilley, C. E., Some aspects of magmatic evolution, *Quart. Jour. Geol. Soc. Lond.* **106**, 37–61 (1950).
276. Tilton, G. R. and Reed, G. W., Radioactive heat production in eclogite and some ultramafic rocks, in *Earth Science and Meteoritics*, 31–43, Ed. J. Geiss and E. D. Goldberg, North-Holland, Amsterdam (1963).
277. Turekian, K. K., The geochemistry of the Atlantic Ocean Basin, *New York Acad. Sci. Trans.* Ser. 2, **26**, 312–30 (1964).
278. Turekian, K. K., and Wedepohl, K. H., Distribution of the elements in some major units of the earth's crust, *Bull. Geol. Soc. Amer.* **72**, 175–92 (1961).
279. Turner, F. J. and Verhoogen, J., *Igneous and Metamorphic Petrology*, 2nd ed., McGraw-Hill, New York (1960).
280. Uchupi, E. and Emery, K. O., The continental slope between San Francisco, California and Cedros Island, Mexico, *Deep-Sea Res.* **10**, 397–447 (1963).
281. Urey, H. C., Lowenstam, H. A., Epstein, S. and McKinney, C. R., Measurement of paleotemperatures and temperatures of the Upper Cretaceous of England, Denmark, and the southeastern United States, *Bull. Geol. Soc. Amer.* **62**, 399–416 (1951).
282. Vacquier, V., Raff, A. D. and Warren, R. E., Horizontal displacements in the floor of the northeastern Pacific Ocean, *Bull. Geol. Soc. Amer.* **72**, 1251–8 (1961).
283. Vine, F. J., Spreading of the Ocean Floor: New Evidence, *Science* **154**, 1405–15 (1966).
284. Vine, F. J. and Matthews, D. H., Magnetic anomalies over oceanic ridges, *Nature* **199**, 947–9 (1963).
285. Vine, F. J. and Wilson, J. T., Magnetic anomalies over a young oceanic ridge off Vancouver Island, *Science* **150**, 485–9 (1965).
286. Wager, L. R. and Deer, W. A., Geological investigations in east Greenland. Pt. 3. The petrology of the Skaergaard intrusion, Kangerdlugssuaq, east Greenland, *Medd. om Grönland* **105**, No. 4 (1939).
287. Wager, L. R., The major element variation in the layered series of the Skaergaard intrusion and a re-estimation of the average composition of the hidden layered series and of the successive residual magmas, *Jour. Petrol.* **1**, 364–98 (1960).
288. Walker, G. P. L., Geology of the Reydarfjördur area, eastern Iceland, *Quart. Jour. Geol. Soc. Lond.* **114**, 367–93 (1958).
289. Walker, G. P. L., The Breiddalur central volcano, eastern Iceland, *Quart. Jour. Geol. Soc. Lond.* **119**, 29–63 (1963).
290. Wangersky, P. J., Manganese in ecology, in *Radioecology*, 499–508, Ed. V. Schultz and A. W. Klement, Reinhold, New York (1963).
291. Wegener, A. L. *The Origin of Continents and Oceans*, Methuen, London (1924).
292. White, W. R. H. and Savage, J. C., A seismic refraction and gravity study of the earth's crust in British Columbia, *Bull. Seismol. Soc. Amer.* **55**, 463–86 (1965).
293. Whitham, K., Measurement of the geomagnetic elements, in *Methods and Techniques in Geophysics* T, 104–67, Ed. S. K. Runcorn, Interscience, New York (1960).
294. Wiens, H. J., *Atoll Environment and Ecology*, Yale University Press, New Haven, Conn. (1962).
295. Williams, H., The Appalachians in northeastern Newfoundland—a two-sided symmetrical system, *Amer. Jour. Sci.* **262**, 1137–58 (1964).
296. Wilson, H. D. B., Andrews, P., Moxham, R. L. and Ramlal, K., Archaean volcanism in the Canadian Shield, *Can. Jour. Earth Sci.* **2**, 161–75 (1965).
297. Wilson, J. T., Cabot Fault, an Appalachian equivalent of the San Andreas and Great Glen Faults and some implications for continental displacement, *Nature* **195**, 135–8 (1962).
298. Wilson, J. T., *A Resumé of the Geology of Islands in the Main Ocean Basins*, Institute of Earth Sciences University of Toronto, Toronto (1963).
299. Wilson, J. T., Continental drift, *Scientific American*, **208**, No. 4, 86–100 (April 1963).
300. Wilson, J. T., Hypothesis of earth's behaviour, *Nature*, **198**, 925–9 (1963).
301. Wilson, J. T., A new class of faults and their bearing on continental drift, *Nature* **207**, 343–7 (1965).

302. Wilson, J. T., Submarine fracture zones, aseismic ridges and the International Council of Scientific Unions Line: proposed western margin of the East Pacific Ridge, *Nature* **207**, 907–11 (1965).

303. Wilson, J. T., Evidence from ocean islands suggesting movement in the earth, in *A Symposium on Continental Drift*, Ed. P. M. S. Blackett, E. Bullard and S. K. Runcorn, *Phil. Trans. Roy. Soc. Lond.* A**258**, 145–67 (1965).

304. Wilson, J. T., Transform faults, oceanic ridges, and magnetic anomalies southwest of Vancouver Island, *Science* **150**, 482–5 (1965).

305. Wolfe, W. J., The Blue River ultramafic intrusion, Cassiar District, British Columbia, *Geol. Surv. Can.* **Paper 64–48** (1965).

306. Wood, J. A., Physics and chemistry of meteorites, in *The Solar System* **4**, 337–401, Ed. B. M. Middlehurst and G. P. Kuiper, University of Chicago Press, Chicago (1963).

307. Wuenschel, P. C., Dispersive body waves — an experimental study, *Geophysics* **30**, 539–51 (1965).

308. Wyllie, P. J., The nature of the Mohorovičić discontinuity, a compromise, *Jour. Geophys. Res.* **68**, 4611–19 (1963).

309. Yoder, H. S. and Tilley, C. E., Origin of basalt magmas: an experimental study of natural and synthetic rock systems, *Jour. Petrol.* **3**, 342–532 (1962).

310. Yonge, C. M., The biology of reef-building corals, *Brit. Mus. (Nat. Hist.) Great Barrier Reef Expedition 1928–1929 Sci. Repts. (London)* **1, No. 13**, 353–91 (1940).

311. Arrhenius, G. and Bonatti, E., Neptunism and vulcanism in the ocean, in *Progress in Oceanography*, Ed. M. Sears and F. Koczy, Pergamon, London (1965).

312. Arrhenius, G., Mero, J. and Korkisch, J., Origin of oceanic manganese minerals, *Science* **144**, 170–3 (1964).

313. Birtill, J. W. and Whiteway, F. E., The application of phased arrays to the analysis of seismic body waves, *Phil. Trans. Roy. Soc.* A **258**, 421–93 (1965).

314. Bolt, B. A. and Turcotte, T., Computer location of local earthquakes within the Berkeley seismographic network, in *Computers in the Mineral Industries*, 561–76, Ed. G. A. Parks, Stanford University (1964).

315. Bonatti, E. and Nayudu, Y. R., The origin of manganese nodules on the ocean floor, *Amer. Jour. Sci.* **263**, 17–39 (1965).

316. Buser, W. and Grutter, A., Über die Natur der Manganknollen, *Schweiz. Mineral Petrog. Mitt.* **36**, 49–62 (1956).

317. Cann, J. R. and Vine, F. J., An area on the crest of the Carlsberg Ridge: petrology and magnetic survey, *Phil. Trans. Roy. Soc.* A **259**, 198–217 (1966).

318. Craig, B. G. and Fyles, J. G., Pleistocene geology of Arctic Canada, *Geol. Surv. Can.* **Paper 60–10** (1960).

319. Curray, J. R., Structure of the continental margin off central California, *Trans. New York Acad. Sci. II* **27**, 794–801 (1965).

320. Degens, E. and Epstein, S., Oxygen and carbon isotope ratios in coexisting calcites and dolomites, *Geochim. et Cosmochim. Acta.* **28**, 23–44 (1964).

321. Fairbridge, R. W., Eustatic changes in sea level, in *Physics and Chemistry of the Earth* **4**, 99–185, Pergamon, London (1961).

322. Fortier, Y. O. and Morley, L. W., Geological unity of the Arctic Islands, *Trans. Roy. Soc. Can.* **50**, (1959).

323. Goldberg, E. D. Chemical and mineralogical aspects of deep-sea sediments, in *Physics and Chemistry of the Earth* **4**, 281–302, Pergamon, London (1961).

324. Eiby, G. A., The New Zealand sub-crustal rift, *N. Z. Jour. Geol. and Geophys.* **7**, 109–33 (1964).

325. Goldschmidt, V. M., *Geochemistry*, Clarendon Press, Oxford (1958).

326. Green, J., Geochemical table of the elements for 1959, *Bull. Geol. Soc. Amer.* **70**, 1127–84 (1959).

327. Haskin, L. A. and Frey, F. A., Dispersed and not-so-rare earths, *Science* **152**, 299–314 (1966).

328. Heezen, B. C., Hollister, C. D. and Ruddiman, W. F., Shaping of the continental rise by deep geostrophic contour currents, *Science* **152**, 502–8 (1966).

329. Herzen, R. P. Von and Langseth, M. G., Present status of oceanic heat-flow measurements, in *Physics and Chemistry of the Earth* **6**, 365–407, Pergamon, London (1965).
330. Marlowe, J. I., Mineralogy as an indicator of long-term current fluctuations in Baffin Bay, *Can. J. Earth Sci.* **3**, 191–201 (1966).
331. Menard, H. W., Sea floor relief and mantle convection, in *Physics and Chemistry of the Earth* **6**, 315–406, Pergamon, London (1965).
332. Pelletier, B. R., Development of submarine physiography in the Canadian Arctic and its relationship to crustal movements, *Bedford Institute of Oceanography*, Manuscript report **64–16** (1964).
333. Richards, F. A., Some current aspects of chemical oceanography, in *Physics and Chemistry of the Earth* **2**, 77–128, Pergamon, London (1957).
334. Talwani, M. and Heirtzler, J. R., Computation of magnetic anomalies caused by two-dimensional structures of arbitrary shape, in *Computers in the Mineral Industries*, 464–80, Ed. G. A. Parks, Stanford, Calif., School of Earth Sciences, Stanford University (1964).
335. Taylor, S. R., The application of trace element data to problems in petrology, in *Physics and Chemistry of the Earth* **6**, 133–213, Pergamon, London (1965).
336. Wangersky, P. J., Sedimentation in three carbonate cores, *Jour. Geol.* **70**, 364–75 (1965).
337. Wangersky, P. J. and Gordon, D. C., Jr., Particulate carbonate organic carbon and Mn^{++} in the open ocean, *Limn. and Oceanog.* **10**, 544–50 (1965).
338. Worzel, J. L., *Pendulum Gravity Measurements at Sea 1936–1959*, Wiley, New York (1965).
339. Hays, J. D. and Opdyke, N. D., Antarctic radiolaria, magnetic reversals and climatic change, *Science* **158**, 1001–11 (1967).
340. Opdyke, N. D., Glass, B., Hays, J. D. and Foster, J., Palaeomagnetic study of Antarctic deep-sea cores, *Science* **154**, 349–57 (1966).
341. Aumento, F., Magmatic evolution of the Mid-Atlantic Ridge, *Earth and Planetary Science Letters* **2**, 225–30 (1967).
342. Shackleton, N., Oxygen isotope analyses and Pleistocene temperatures re-assessed, *Nature* **215**, 15–17 (1967)

Index